# Renormalized Perturbation Theory and its Optimization by the Principle of Minimal Sensitivity

# Renormalized Perturbation Theory and its Optimization by the Principle of Minimal Sensitivity

**P M Stevenson**
*Rice University, USA*

 **World Scientific**

NEW JERSEY · LONDON · SINGAPORE · BEIJING · SHANGHAI · HONG KONG · TAIPEI · CHENNAI · TOKYO

*Published by*

World Scientific Publishing Co. Pte. Ltd.

5 Toh Tuck Link, Singapore 596224

*USA office:* 27 Warren Street, Suite 401-402, Hackensack, NJ 07601

*UK office:* 57 Shelton Street, Covent Garden, London WC2H 9HE

**British Library Cataloguing-in-Publication Data**

A catalogue record for this book is available from the British Library.

ISBN 978-981-125-568-7 (hardcover)
ISBN 978-981-125-569-4 (ebook for institutions)
ISBN 978-981-125-570-0 (ebook for individuals)

For any available supplementary material, please visit
https://www.worldscientific.com/worldscibooks/10.1142/12817#t=suppl

Desk Editor: Nur Syarfeena Binte Mohd Fauzi

Typeset by Stallion Press
Email: enquiries@stallionpress.com

# Acknowledgments

The research on which this book is based was funded largely by the U.S. Department of Energy through grants to the University of Wisconsin-Madison and to Rice University. The author is most grateful for that support. Some work, of 1981–82, was funded by a CERN Fellowship.

# Contents

# Chapter 1

# Introduction

## 1.1. Prologue and Preview

With renormalized perturbation theory we are not dealing with an ordinary power series, but with a more subtle mathematical structure. A renormalized series, such as

$$\mathcal{R} = a(1 + r_1 a + r_2 a^2 + \cdots),\tag{1.1}$$

is an expansion in a parameter $a$ that is indefinite, in that there are infinitely many "$a$'s," all fundamentally on an equal footing. They are related by transformations of the form

$$a' = a(1 + v_1 a + v_2 a^2 + \cdots)\tag{1.2}$$

with arbitrary finite coefficients $v_1, v_2, \ldots$. The series coefficients are correspondingly indefinite; they transform as

$$r_1' = r_1 - v_1,$$
$$r_2' = r_2 - 2r_1 v_1 + 2v_1^2 - v_2,\tag{1.3}$$

$$\vdots$$

so that $\mathcal{R}$

$$\mathcal{R} = a'(1 + r_1' a' + r_2' a'^2 + \cdots)\tag{1.4}$$

remains invariant.

The invariance of $\mathcal{R}$ is called "renormalization-group (RG) invariance." The essential point is that renormalization — the procedure for handling ultraviolet divergences in quantum field theory (QFT) — introduces a "renormalized couplant" $a$, whose precise definition — the "renormalization scheme" (RS) — involves a great deal of arbitrariness. Changing from one RS to another induces transformations of the form (1.2). However, any physically measurable quantity $\mathcal{R}$ must be RG invariant: that is, independent of the RS used.

By itself, the mathematical structure just described is too "floppy" to be of much interest. It is the physical consideration of *Dimensional Analysis* that gives the problem a backbone. Any RS involves choosing a "renormalization scale" $\mu$, which is an arbitrary parameter with dimensions of mass. The dependence of $a$ on $\mu$ is given by the famous $\beta$ function:

$$\mu \frac{da}{d\mu} \equiv \beta(a) = -ba^2(1 + ca + c_2 a^2 + c_3 a^3 + \cdots). \qquad (1.5)$$

The $\beta$ function plays a key role and is needed to connect $a$ to the free parameter of the theory (a suitably defined $\Lambda$ parameter). The first two coefficients, $b$ and $c$, are numbers that can be unambiguously calculated in any given QFT. It is straightforward to show (see Sec. 6.2) that both are invariant under the transformations of Eq. (1.2). The higher coefficients $c_2, c_3, \ldots$ can also be calculated, but their values depend on the RS choice. The coefficients $r_1, r_2, \ldots$ in the perturbative expansion of any specific physical quantity $\mathcal{R}$ can also be calculated (in the same RS). Although the $r_i$'s and the $c_j$'s are separately RS dependent, there are certain combinations of them that are RS invariants. By forming these invariants one distils the RG-invariant, physical content of the Feynman-diagram calculations from the scheme artefacts.

Beyond the formal level one must define, at a given order, the analog of the "partial sum." In a given RS that involves truncating both the $\mathcal{R}$ and $\beta$ series at the same order. However, the approximant thus defined is RS dependent, creating the notorious *RS-dependence problem*. Standard QFT textbooks typically mention the issue, and then mostly ignore it. Yet, without a resolution of the problem, the results of perturbative QFT will remain unclear and ambiguous. (The complacent assumption that the difficulty becomes less and less severe at higher orders is unsafe because the fixed-scheme series are divergent.) The problem, it should be stressed, is not with the theory itself, but arises from the approximation.

The approach in this book is founded on the *Principle of Minimal Sensitivity*, which is a philosophy that applies quite generally to any "non-invariant approximation." In such situations the exact result is known to be independent of certain "extraneous parameters," yet the approximate result depends upon them. The idea is to choose the values of the extraneous parameters so that the approximant is minimally sensitive to small variations of those values. Thus, an exact symmetry of the exact result is mimicked by a local symmetry of the approximate result. Applied to RS dependence, this principle identifies an "optimal" result that is determined, for a given physical quantity at a given order, by a set of "optimization equations." From those equations we can solve for the optimized $r_i$ coefficients in terms of $a$ and the $c_j$ coefficients. If we know, from Feynman diagram calculations, the numerical values of the invariants to that order, then we may obtain the optimized result for $\mathcal{R}$ numerically, by a suitable iterative procedure.

Optimization has interesting consequences, particularly for the infrared limit, and also for the question of high-order behaviour, where we conjecture that optimization may produce a convergent sequence of approximations, despite the generic divergence of QFT series in any fixed RS.

## 1.2. Plan of the Book

The core of the book is Part II, a thorough exposition of "optimized perturbation theory" (OPT). The impatient reader, already familiar with QFT and renormalization, may wish to begin there. However, the chapters making up Part I provide background material intended to set the RS problem in context and to explain the thinking behind this approach. It is hoped that students of QFT will find these chapters helpful in clarifying some key conceptual issues in the RG and renormalized perturbation theory; issues that can sometimes be obscured by technical details in textbooks. The chapters on non-invariant approximations and induced convergence are relevant to a huge variety of problems in all areas of physics and applied mathematics.

> **Parenthetical remarks,** such as this one, are interspersed in the text in just this format. They may often be ignored without affecting the reader's understanding of the main line of argument. They discuss more technical issues, or address questions, complications, or apparent difficulties that might occur to some readers at that point.

Chapter 2 discusses Dimensional Analysis in QFT, which is the key to understanding how a seemingly dimensionless bare coupling constant turns into an "effective" or "running" coupling constant that varies with energy. It also explains much about why RG equations arise.

Chapter 3 discusses renormalization in conceptual terms, suppressing all technical details. It emphasizes that coupling-constant renormalization, for physical quantities, is essentially a substitution, a reparametrization of the theory. The ill-defined (infinitesimal) bare coupling constant is eliminated in favour of a finite renormalized couplant, thereby re-organizing the perturbation series into a usable form. However, the freedom inherent in defining a renormalized couplant leads, at any finite order, to the RS ambiguity.

This RS-dependence problem arises only because we *truncate* the perturbation series. It is not fundamentally a QFT issue, but rather is a problem in "approximology" — the theory and lore of approximations. We therefore open Chapter 4 with a general discussion of approximations, stressing that they involve taking a gamble based on incomplete information. A good gambler, however, always makes wise use of all available information. For "non-invariant approximations," the vital extra information is the knowledge that the exact result does not depend on certain variables. The *Principle of Minimal Sensitivity*, which makes use of this information, is explained and applied to several examples, in particular the Caswell–Killingbeck (CK) expansion for the anharmonic oscillator in quantum mechanics.

Chapter 5 discusses the phenomenon of "induced convergence" that can occur in high orders of non-invariant approximations. As well as the CK expansion, it discusses a toy model involving the alternating factorial series, where the convergence of the optimized results can be proved quite simply. We conjecture that OPT in quantum field theories will also show such induced convergence, even though the $\mathcal{R}$ and $\beta$ series, in a fixed RS, are factorially divergent.

In Part II the main work begins. Chapters 6 and 7 set out the formal structure of renormalized perturbation series. They explain how to parametrize RS dependence and identify the invariants $\rho_i$ formed from the coefficients of $\mathcal{R}$ and $\beta(a)$. The discussion, unfortunately, has to begin with $\mu$ dependence, which arises at the lowest order, but which is, in some ways, the most difficult and confusing issue. Thus, it is first necessary to discuss several preliminaries in Chapter 6: the $\beta$ function; its integration; the appropriate definition of a boundary-condition parameter $\tilde{\Lambda}$; and how $\tilde{\Lambda}$ depends on RS. Only then can one see that the first of the RS variables is not $\mu$ itself, but the *ratio* of $\mu$ to $\tilde{\Lambda}$. A special energy-dependent invariant, $\boldsymbol{\rho}_1(Q)$, is involved here. The way is then clear to discuss, in Chapter 7, the other RS dependences that can be parametrized by the higher-order $\beta$-function coefficients $c_j$ for $j = 2, 3, \ldots$. Here things are in some ways more straightforward. A set of invariant quantities $\rho_j$ can be defined as specific combinations of the coefficients of $\mathcal{R}$ and $\beta(a)$. Only at the end can one see how the $\mu/\tilde{\Lambda}$ dependence corresponds to the "$j = 1$" case of the general formulas.

**It is as if** a teacher, introducing integral calculus for the first time, were compelled to first discuss $\int \frac{dx}{x}$, necessitating a lengthy discussion of the natural logarithm function, $\ln x$, before turning to the integrals of positive integer powers and deducing the simple formula $\int dx\, x^{n-1} = x^n/n$. Only then, by considering $n = \epsilon$, could the $\int \frac{dx}{x}$ result be seen as part of the general pattern — the subtlety being that $x^\epsilon \sim 1 + \epsilon \ln x$ as $\epsilon \to 0$.

Chapter 8 addresses the issue of finite orders, where series are truncated and exact RS invariance is lost. It explains how to apply the *Principle of Minimal Sensitivity*, from which all later results follow. Chapter 9 shows that the resulting optimization equations can be solved to yield the optimized $r_m$ coefficients in terms of the optimized $c_j$ coefficients and the optimized couplant $a$. It also outlines algorithms for numerically solving the remaining equations iteratively. Chapter 10 presents in detail some illustrative numerical results for the $R_{e^+e^-}$ ratio in QCD.

Part III discusses various special topics. Chapter 11 explores the infrared limit, where perturbation theory, for an asymptotically free theory, is furthest from its comfort zone. Chapter 12 treats the important topic of factorized quantities, such as moments of the proton structure function, where one factor, the operator matrix element, is not perturbatively calculable. Chapter 13 is of more specialized interest. It considers QCD in the small-$b$ limit (known as the Banks–Zaks or BZ limit) where one can explore some aspects of all-orders OPT. Although not addressing the main issue concerning the "induced convergence" conjecture, because the fixed-RS series in this limit is not divergent, these explorations do shed some light on how OPT works at very high orders.

Some issues, falling outside the main narrative, are discussed in appendices to some of the chapters. A few end-of-chapter exercises are included; these are mainly intended to enable readers to test and reinforce their understanding of the material, but sometimes contain results of interest, though not of central importance. References have been kept to a bare minimum in the text, though names are mentioned where important credit is due. More extensive citations of the prior and contemporaneous literature can be found in the author's research papers. These are listed in the bibliography along with other signposts to the literature. The bibliography also mentions a selection of other applications of the minimal-sensitivity criterion; in particular some promising developments in QCD that go beyond the scope of this book.

## 1.3. Quantum Chromodynamics

The principal application of OPT is to quantum chromodynamics (QCD) so we will very briefly describe the theory. This is only intended to give readers without much particle-physics background a general feeling for the theory — we mainly want to show how the gauge coupling $g$ (and hence the couplant $a \equiv g^2/(4\pi^2)$) enters. Readers are advised to allow the technical details to "flow over their heads" if necessary, or to consult other textbooks for a fuller description.

QCD is a relativistic quantum field theory. Being *relativistic*, it is natural to use units where the speed of light is unity, and to use covariant notation. 4-vectors, such as the space–time position $x^\mu = (t, \boldsymbol{x})$ and energy–momentum $p^\mu = (E, \boldsymbol{p})$, have a Lorentz index $\mu = 0, 1, 2, 3$ that is lowered with the Minkowski metric tensor $g_{\mu\nu} = \mathrm{diag}(1, -1, -1, -1)$. An invariant is formed by $x^\mu x_\mu = x^\mu g_{\mu\nu} x^\nu$ (with summation over repeated indices always understood). The upper or lower index distinction is crucial for Lorentz indices.

Being a *field* theory, QCD's degrees of freedom are fields — which, at the classical level, are just functions of space–time position. The simplest kind of field is a scalar field $\phi(x)$, and the simplest theory, describing free, spinless particles of mass $m$, has the Lagrangian density

$$\mathcal{L} = \frac{1}{2} \partial^\mu \phi \, \partial_\mu \phi - \frac{1}{2} m^2 \phi^2. \qquad (1.6)$$

It can be thought of as the continuum limit of a set of oscillators, one at each spatial point, each coupled to its nearest neighbours.

Being a *quantum* theory, the fields are upgraded to operators, with quantum-oscillator commutation relations. Alternatively, one can formulate the quantum theory in terms of a functional integral, over all possible classical field configurations, involving the exponential of the action, $\int d^4x \mathcal{L}$.

QCD involves spin-$\frac{1}{2}$ particles (quarks) and spin-1 particles (gluons). Particles with spin are described by multi-component fields. Spatial rotations and Lorentz boosts result in a mixing of the components (in a way that we will not attempt to describe here). A spin-1 field $A^\mu(x)$ has four real components, labelled by a Lorentz index $\mu$ that can take the values $0, 1, 2, 3$. A spin-$\frac{1}{2}$ field $\psi_\alpha(x)$ has four complex components, labelled by a Dirac index $\alpha$.

The Lagrangian density for a free spin-$\frac{1}{2}$ fermion is

$$\mathcal{L} = \bar{\psi}_\alpha \left( i(\gamma^\mu)_{\alpha\beta}\,\partial_\mu - m\,\delta_{\alpha\beta} \right) \psi_\beta. \tag{1.7}$$

Here $\gamma^\mu$ are the four Dirac matrices, satisfying an anticommutation relation $\{\gamma^\mu, \gamma^\nu\} = 2g^{\mu\nu}$. Henceforth we will suppress the Dirac indices entirely, leaving it understood that a matrix multiplication is involved. The shorthand notation $\slashed{\partial}$ for $\gamma^\mu \partial_\mu$ is also convenient, so we have

$$\mathcal{L} = \bar{\psi} \left( i\slashed{\partial} - m \right) \psi. \tag{1.8}$$

$\bar{\psi}$ is the adjoint of $\psi$ (Hermitian conjugate times the $\gamma^0$ Dirac matrix).

This Lagrangian is obviously invariant under a change of phase from $\psi$ to $e^{i\theta}\psi$, since $\bar{\psi}$ transforms to $e^{-i\theta}\bar{\psi}$. Here $\theta$ can be any constant. Suppose, however, that we want to have a *gauge* symmetry where $\theta$ can be a function of space–time position $x$. That can be achieved if we introduce a "covariant derivative"

$$D^\mu = \partial^\mu + igA^\mu \tag{1.9}$$

involving a new vector field $A^\mu$ that transforms as

$$A^\mu \longrightarrow A^\mu - \frac{1}{g}\partial^\mu\theta. \tag{1.10}$$

In order for the $A_\mu$ field to be dynamical we need to add a kinetic term for it. Naïvely, this would be something like $\partial_\nu A^\mu \partial^\nu A_\mu$, akin to the kinetic term in Eq. (1.6), but that would not be invariant under the gauge transformation (1.10). The appropriate term can be found by considering the commutator of two covariant derivative operators:

$$[D^\mu, D^\nu] = ig(\partial^\mu A^\nu - \partial^\nu A^\mu) \equiv igF^{\mu\nu} \tag{1.11}$$

so that the appropriate gauge-invariant kinetic term is $-\frac{1}{4}F_{\mu\nu}F^{\mu\nu}$. If we identify the fields $A_\mu$ and $\psi$ with the photon and electron, respectively, and the gauge coupling with the electric charge $e$, then this is Quantum Electrodynamics (QED):

$$\mathcal{L}_{QED} = -\frac{1}{4}F_{\mu\nu}F^{\mu\nu} + \bar{\psi}\left(i\slashed{D} - m\right)\psi. \tag{1.12}$$

Unlike the previous Lagrangians, which were quadratic in the fields and so were theories of non-interacting particles, the new Lagrangian has a cubic term coming from the $\bar\psi \not{D}\psi$ term, which is

$$- gA_\mu\bar\psi\gamma^\mu\psi. \tag{1.13}$$

This term, treated as a perturbation to the free-electron and free-photon theories, corresponds to a vertex in the Feynman diagrams where a photon line meets up with an incoming and an outgoing electron line.

To get QCD we start from a "quark" field, a threefold fermion field $\psi_a$, where $a$ is the "colour" index; $a = 1, 2, 3$ (or $a =$red, green, blue). Summing over colours, as in $\bar\psi_a\psi_a$, produces a "colourless" combination. Crucially, all three fields have the same mass so that the Lagrangian

$$\bar\psi_a(i\not\partial - m)\psi_a \tag{1.14}$$

has an exact symmetry under $\psi_a \longrightarrow \psi'_a$, with

$$\psi'_a = U_{ab}\psi_b, \tag{1.15}$$

where $U_{ab}$ is any $3 \times 3$ unitary matrix with unit determinant. This symmetry is called SU(3) colour symmetry. (There is also a symmetry under an overall $e^{i\theta}$ phase factor, which gives rise to electromagnetic interactions of the quark with the photon field.) A continuous group, such as SU(3), is characterized by its "Lie algebra," the set of commutators of its generators $T^A$:

$$[T^A, T^B] = if_{ABC}T^C. \tag{1.16}$$

For SU(3) the index $A$ runs from 1 to 8 ($= 3^2 - 1$) and is raised or lowered with a simple Kroenecker delta $\delta_{AB}$. The group's structure constants, $f_{ABC}$, are totally antisymmetric in the three indices. The individual elements are pure numbers. ($f_{123} = 1$, $f_{458} = f_{678} = \sqrt{3}/2$, $f_{147} = f_{165} = f_{246} = f_{257} = f_{345} = f_{376} = \frac{1}{2}$, with the others fixed by antisymmetry, or otherwise zero.) The generators should be thought of as abstract objects, but there are various matrix *representations* of the Lie algebra. The relevant ones here are the fundamental representation, where the $T^A$'s are represented by

$3 \times 3$ Hermitian matrices $T^A_{ab}$, and the adjoint representation, where they are represented by $8 \times 8$ matrices $T^A_{BC}$ whose matrix elements are directly given by $-if_{ABC}$'s. A general $U_{ab}$ matrix can be written as

$$U_{ab} = \exp(i\theta_A T^A_{ab}) \tag{1.17}$$

with eight parameters $\theta_A$ multiplying the eight $T^A_{ab}$ matrices.

Again, we wish to make this symmetry into a gauge symmetry where the $\theta_A$'s can be functions of $x$. We need a covariant derivative

$$D^\mu = \partial^\mu + igT^A G^\mu_A \tag{1.18}$$

involving an eightfold gauge field $G^\mu_A$ (the "gluon" field) that transforms as

$$G^\mu_A \longrightarrow G^\mu_A - f_{ABC}\theta^B G^{C,\mu} - \frac{1}{g}\partial^\mu \theta_A \tag{1.19}$$

The $f_{ABC}$ term is the natural one for the transformation of an object with an adjoint index $A$; the additional inhomogeneous term makes $G^\mu_A$ a gauge field.

Again, we need a kinetic term for this gauge field. To find the right gauge invariant form, we again consider the commutator of two covariant derivative operators:

$$[D^\mu, D^\nu] = igT^A \left(\partial^\mu G^\nu_A - \partial^\nu G^\mu_A - gf_{ABC}G^{B,\mu}G^{C,\nu}\right). \tag{1.20}$$

Note that the commutator here involves two sorts of non-commuting objects; derivative operators and generators, with the latter giving rise to the $gfG^2$ term. The analogue of the $F^{\mu\nu}$ tensor in QED is thus

$$\mathcal{G}^{\mu\nu}_A \equiv \partial^\mu G^\nu_A - \partial^\nu G^\mu_A - gf_{ABC}G^{B,\mu}G^{C,\nu}. \tag{1.21}$$

This object transforms in the "proper" way for an object with an $A$ index — like Eq. (1.19) but *without* the derivative term. Thus, when we form the kinetic term,

$$-\frac{1}{4}\mathcal{G}^A_{\mu\nu}\mathcal{G}_A{}^{\mu\nu}, \tag{1.22}$$

it is gauge invariant.

There are various different "flavours" of quarks, so we need yet another index $j$ to label them that is summed from 1 to $n_f$, the total number of flavours. In the real world, we know of six flavours called $u, d, s, c, b, t$. Each flavour has a different mass. The Lagrangian for QCD is then

$$\mathcal{L}_{\text{QCD}} = -\frac{1}{4}\mathcal{G}^A_{\mu\nu}\mathcal{G}_A{}^{\mu\nu} + \sum_{j=1}^{n_f} \bar{\psi}_{j,a}(i\not{D}_{ab} - m_j\delta_{ab})\psi_{b,j}. \qquad (1.23)$$

(There are more technicalities about gauge fixing, Fadeev–Popov ghosts, and BRST invariance that we do not go into here, though they are crucial for proving the theory's renormalizability.)

Besides the quark–quark–gluon interaction term of the form $-g(\bar{\psi}\gamma^\mu\psi)G$ from the $\not{D}$ term, there are also three-gluon and four-gluon interaction terms $g(\partial_\mu G)G^2$ and $g^2 G^4$. These give rise to corresponding vertices in the Feynman diagrams.

Feynman-diagram calculations, especially at higher orders, are much more complicated when masses are included, so it is usual to approximate "real QCD" by massless QCD with an effective number of "active" flavours. (Quarks with masses above the relevant energy scale $Q$ are regarded as "inactive" and the lighter quarks are treated as massless.)

Other gauge theories can be constructed in a similar fashion, starting with a different symmetry group, such as SU($N$) instead of SU(3). In all cases, the gauge coupling $g$ is naturally dimensionless in 4 dimensions.

# Part I

# Renormalization and Its Ambiguities

## Chapter 2

# Dimensional Analysis
# in Quantum Field Theory

## 2.1. Physics and Conventions

A quantitative description of a physical process always involves a
convolution of two parts: a **Ph** part — the physics, the phenomenon,
the "phacts" — and a **C** part — the conventions, choices of units,
coordinate system, etc., chosen (by committees) for concreteness,
convenience, and clear communication:

$$
\boxed{\begin{array}{c}\text{Quantitative}\\\text{description of a}\\\text{physical process}\end{array}} = \boxed{\textbf{Physics}} \otimes \boxed{\text{Conventions}} \ .
$$

All the messiness and arbitrariness lies in the **C** part, while the
beauty resides in the **Ph** part. The actual physics is fundamentally
independent of the arbitrary choices in the **C** part. Thus, for instance,
Articles **I** and **II** of the Physicist's Creed are

**I.** Physics is independent of our choice of units.
**II.** Physics is independent of our choice of coordinate system.

These points, though elementary, are nevertheless profound.
Article **I** is fundamental to our discussion of renormalization-group
equations here. Article **II**, for non-inertial space–time coordinate
systems, leads towards General Relativity. As physics advances,

it is striking that our descriptions tend to involve ever more sources of redundancy: for instance, physics is independent of our choice of Lorentz frame, our choice of gauge — and our choice of renormalization scheme.

Theoretical work can often be done in mathematical formalisms where these invariances are manifest. However, when it comes to making concrete, quantitative predictions that can be compared with experiment, we are forced to adopt some definite system of units, coordinate system, etc. In doing so it is vital that we do not lose sight of the fundamental invariance of the physics to those arbitrary parameters.

## 2.2. Dimensional Analysis

Let us focus now on Article **I**: — physics is independent of our choice of units. Most quantities in physics are not pure numbers. When we describe a length as "3.2 metres" we mean that the *ratio* of that length to another specified length, previously defined to be "1 metre," is the number 3.2. Generally, quantities in physics have *dimensions* and must be expressed in some well-defined *units* before they can be identified with *numbers*. Traditionally the fundamental dimensions are taken to be *mass, length, time*, with the dimensions of other quantities being taken as combinations of these. (For instance, angular momentum has dimensions $(mass)(length)^2(time)^{-1}$.) Dimensional Analysis (DA) is based on the fact that it only makes sense to add, subtract, or equate two terms if they have the *same* dimensions. This simple observation — as any good physics student appreciates — provides a powerful check against mistakes in calculations. It also provides a way to predict, in advance, much about the form that a result must take. DA applies to QFT too, of course. In some ways it is even simpler, though in other ways it is more subtle.

DA in QFT is simpler because the fundamental constants $\hbar$ (Planck's constant over $2\pi$) and $c$ (the speed of light in vacuum) are ubiquitous and provide natural units of angular momentum and velocity, respectively. Thus, we may reduce the three fundamental dimensions to just one, which may be taken to be *mass* (or, equivalently, *energy*, or *inverse length*, etc.) and all quantities can be regarded as having dimensions of mass to some power. We will use the phrase "a massive parameter," to mean "a parameter with the dimensions of mass."

DA in renormalizable QFT's is more subtle because of the need for renormalization. This subtlety shows up most clearly in *massless* (or, more precisely, classically scale-invariant) theories through the counter-intuitive phenomenon of "dimensional transmutation." The bare Lagrangian of a *massless* renormalizable theory contains no parameters with dimensions, only a dimensionless bare coupling constant $g$; nevertheless, the theory ends up being characterized by a scale parameter "$\Lambda$" with dimensions of mass. The aim of this chapter is to explain how dimensional transmutation is consistent with DA. We will show that much, though not all, the content of the RG equations presented in QFT textbooks follows simply from DA. In the following discussion, the theory is assumed to be a massless, renormalizable QFT with a single bare coupling constant, $g$, with massless QCD as a specific example. (Appendix 2.A comments briefly on theories with masses.)

## 2.3. The Dimensional Transmutation "Paradox"

For the purposes of discussion, let us picture our theory as a "Black Box" that provides the answer to any well-posed, physical question that we put to it. This device — alas, rhetorical not actual — is employed in order to separate, as far as possible, the theory in itself and the way *it* works, from the way in which *we* work the theory. Later on we will return to the complications that arise because we are forced to use approximate methods — indeed, those complications are the main subject of this book. For the present, however, let us pretend that we have a "Black Box" that can solve the theory exactly. The "paradox" of dimensional transmutation can then be described in the following terms.

The Lagrangian we input to the Black Box is scale-invariant — it contains no massive constants. Let us ask the theory to give its prediction for $\mathcal{A}(Q)$, a specific, dimensionless, observable quantity, which can depend only on one *massive* variable $Q$. For example, in QCD we could ask for the dimensionless ratio of cross sections

$R_{e^+e^-}(Q) \equiv \sigma_{\text{tot}}(e^+e^- \to \text{hadrons})/\sigma(e^+e^- \to \mu^+\mu^-)$ at a center-of-mass energy $Q$. The answer from the Black Box turns out to have a non-trivial dependence on $Q$. In QCD, $R_{e^+e^-}(Q)$ tends to a constant only as $Q \to \infty$; there are definite sub-asymptotic terms that vary with $Q$.

Finding that the dimensionless $\mathcal{A}(Q)$ actually depends on $Q$ is quite a shock because we know that the following is true:

**Theorem (Dimensional Analysis).** *A function $f(x, y)$ which depends only on two massive variables $x$ and $y$, and which is*

(i)  *dimensionless,*
(ii)  *uniquely defined,*
(iii)  *whose definition does not involve any massive constants*

*must be a function of the ratio $x/y$ only.*

**Corollary.** *If $f(x, y)$ is independent of $y$, then it must be a constant.*

The question we have asked of the theory is of the type covered by the corollary: the theory has to define a dimensionless function of one massive variable, and it has no massive constants available to it. Thus, the answer to our question *must* be a constant — and yet, the answer from the Black Box is certainly *not* a constant!

The flaw in the "paradox" is the requirement that the function must be *uniquely* defined in the theorem above. If this requirement is dropped, then requirement (iii) can easily be circumvented by allowing a massive constant to appear, *implicitly*, as a constant of integration. The point is that the Lagrangian we input is not one single theory but a one-parameter set of theories with different (bare) coupling constants, $g$. Therefore, our Black Box gives us not just *one* answer, but a one-parameter set of answers — that is, it specifies the dimensionless function $\mathcal{A}(Q)$ *non-uniquely*. The paradox is then resolved: the DA theorem is correct, but it does not apply, because the uniqueness requirement is not satisfied.

Dimensional Transmutation is the process whereby a theory exploits its one-parameter ambiguity so as to evade the dull consequences of the DA theorem. In order to use this mechanism, a theory must (i) have at least one free parameter, $g$, and (ii) have infinities (or be ill-defined in some other way) so that its predictions cannot be expressed directly in terms of $g$. In field theories, the second requirement is met by the necessity for renormalization. A field-theoretic Lagrangian is not a complete description of the theory: in order to make sense of it one must use some renormalization procedure. The physics is independent of all the details of the renormalization prescription, but not of the *necessity* for renormalization.

*To summarize:* we asked the Black Box a question: "What is $\mathcal{A}(Q)$?", naïvely expecting that the answer would be a constant. Having now understood that the theory will give us only a one-parameter ambiguous answer, we can reconsider what DA tells us about the form that this answer must take.

## 2.4. The Ambiguous Answer for $\mathcal{A}(Q)$

The Black Box is asked to specify $\mathcal{A}(Q)$ and — while it *is* allowed a one-parameter ambiguity — there are no massive constants it can use. So, in order to produce a non-constant $\mathcal{A}(Q)$, the theory's definition must be somewhat indirect. One can immediately think of two possibilities: the black box could specify only the first derivative of $\mathcal{A}(Q)$, or it could define $\mathcal{A}(Q)$ recursively. These two alternatives are actually equivalent, as we shall show explicitly.

If the theory specifies the first derivative of $\mathcal{A}(Q)$, then this specification must take the form

$$\frac{d\mathcal{A}}{dQ} = \frac{\mathcal{B}(\mathcal{A})}{Q}, \tag{2.1}$$

where $\mathcal{B}(\mathcal{A})$ is a dimensionless function of a dimensionless argument and must be *uniquely* defined by the theory. The important point is that the *form* of Eq. (2.1) is unique: the explicit $Q$-dependence is fixed by DA.

If instead the specification of $\mathcal{A}(Q)$ is recursive, then it must have the form

$$\mathcal{A}(Q) = F\left(\frac{Q}{\mu}, \mathcal{A}(\mu)\right) \quad \text{for any } \mu, \tag{2.2}$$

where $F$ is a uniquely defined, dimensionless function which must be independent of the arbitrary massive parameter $\mu$, because no massive constants are allowed in the specification of $\mathcal{A}(Q)$. Again, the form of the equation is dictated by DA. The one-parameter ambiguity arises because the theory does not provide a boundary condition for Eq. (2.1) or Eq. (2.2). This boundary condition must be fixed by an appeal to experiment.

Let us first analyze Eq. (2.1). Integrating the equation gives

$$\ln Q + \text{constant} = \int_{\infty}^{\mathcal{A}} \frac{dA}{\mathcal{B}(A)} \equiv K(\mathcal{A}). \qquad (2.3)$$

**For present purposes** a lower limit of $\infty$ is simple and convenient. However, that choice would not be appropriate in perturbation theory and a different definition will be discussed later in Sec. 6.3.

The constant of integration can be written in infinitely many ways, corresponding to different choices of mass units, by writing "constant $= K_0 - \ln \mu$," where $\mu$ is some arbitrary massive parameter. By considering the point $Q = \mu$ one sees that $K_0 = K(\mathcal{A}(\mu))$, so that

$$\ln(Q/\mu) + K(\mathcal{A}(\mu)) = K(\mathcal{A}(Q)). \qquad (2.4)$$

What has really happened is that a *massive constant* has crept in as an integration constant. One way of characterizing this constant, $\Lambda$, is as the value of $\mu$ for which $K(\mathcal{A}(\mu))$ vanishes (assuming that such a point exists). Thus Eq. (2.1) is equivalent to

$$\mathcal{A}(Q) = K^{-1}\left(K(\mathcal{A}(\mu)) + \ln(Q/\mu)\right) \quad \text{for *any* } \mu, \qquad (2.5)$$

or to

$$\mathcal{A}(Q) = K^{-1}\left(\ln(Q/\Lambda)\right) \quad \text{for *some* } \Lambda. \qquad (2.6)$$

Equation (2.5) holds for arbitrary $\mu$, and $\partial\mathcal{A}/\partial\mu|_Q = 0$ for all $Q$, as the reader should verify. Equation (2.6), however, holds only for some fixed, but unspecified constant, $\Lambda$. We could have added "where $\Lambda$ is defined by $K(\mathcal{A}(\Lambda)) = 0$," but this "definition" is mere tautology, since this information is already contained in Eq. (2.6). The unspecified nature of $\Lambda$ reflects the one-parameter ambiguity of the theory.

Next let us investigate Eq. (2.2) to show that it also leads to Eqs. (2.5), (2.6). Changing the variables $Q, \mu$ to

$$z \equiv Q/\mu, \quad A \equiv \mathcal{A}(\mu), \tag{2.7}$$

the condition that $F$ is independent of $\mu$ is

$$\mu \frac{dF}{d\mu} = \frac{\partial F}{\partial z}(z, A) \bigg|_A \mu \frac{dz}{d\mu} + \frac{\partial F}{\partial A}(z, A) \bigg|_z \mu \frac{d\mathcal{A}(\mu)}{d\mu} = 0. \tag{2.8}$$

Since $\mu \frac{dz}{d\mu} = -z$, this becomes

$$z \frac{\partial F}{\partial z}(z, A) \bigg|_A \bigg/ \frac{\partial F}{\partial A}(z, A) \bigg|_z = \mu \frac{d\mathcal{A}(\mu)}{d\mu} \tag{2.9}$$

The right-hand side of this equation is independent of $z$ and is a function of $A$ only. In fact, it is precisely the function $\mathcal{B}(A)$ of Eq. (2.1). Equating the left-hand side of Eq. (2.9) to $\mathcal{B}(A)$ leads immediately to

$$\frac{\partial F}{\partial (\ln z)} = \frac{\partial F}{\partial (K(A))}, \tag{2.10}$$

where

$$K(A) \equiv \int_{\infty}^{A} \frac{dA'}{\mathcal{B}(A')}, \tag{2.11}$$

as before. The solution to Eq. (2.10) is of course that $F$ can depend only on the single variable $K(A) + \ln z$. Hence, reverting to $Q, \mu$ as the variables,

$$F(Q/\mu, \mathcal{A}(\mu)) = f\left(K(\mathcal{A}(\mu)) + \ln(Q/\mu)\right), \tag{2.12}$$

where $f$ is a function of a single variable. The left-hand side is, by definition, $\mathcal{A}(Q)$, so by considering $\mu = Q$ one sees that $\mathcal{A}(Q) = f(K(\mathcal{A}(Q)))$, so $f$ must be $K^{-1}$. Thus, one recovers Eq. (2.5) as promised. Equations (2.1) and (2.2) are therefore equivalent definitions of $\mathcal{A}(Q)$. In fact, any answer the Black Box could give would have to be equivalent to the form (2.1).

Notice the curious feature that the discussion often involves an arbitrary massive parameter, $\mu$. Its role is that of a conjurer's handkerchief — "Now you see it, now you don't!" It is necessary in order to write down certain formulas, but physically it is never really there.

The massive constant $\Lambda$ is quite different. Its value does have physical significance. The theory does not, however, *predict* its value. In fact, since $\Lambda$ is not in its input Lagrangian, the Black Box cannot ever mention $\Lambda$. Nor does it need to: it simply leaves a boundary condition unspecified. We ourselves introduce $\Lambda$ as a convenient means of parametrizing the theory. By expressing the theory's predictions in terms of $\Lambda$ and then fitting these to $N$ experimental measurements, we can extract the value of $\Lambda$ and make $(N-1)$ tests of the theory.

*To summarize:* There is essentially only one form in which the Black Box can give the answer to our question "What is $\mathcal{A}(Q)$?" It must provide a fully specified function $\mathcal{B}(A)$, where $\mathcal{B}$ and its argument are dimensionless, together with the statement that $\mathcal{A}(Q)$ is related to $\mathcal{B}(\mathcal{A})$ by Eq. (2.1). This is equivalent to Eqs. (2.2), (2.5), (2.6) as shown above. The Black Box cannot ever mention a massive constant in its answers. Nevertheless, we can see that a dimensional constant does creep into the theory because of the absence of a boundary condition.

## 2.5. The Form of Other Physical Quantities

Now that we understand the form of its answer for one specific physical quantity, $\mathcal{A}(Q)$, let us ask the Black Box to predict some other physical quantity. The answer cannot be an independent repeat of the previous story, since this would introduce a second "$\Lambda$" parameter, and there is only a *one*-parameter ambiguity in the theory. Instead, having already given us $\mathcal{A}(Q)$, the Black Box can now give all its other results in terms of $\mathcal{A}$.

Consider some observable quantity $\sigma(q_1, \ldots, q_n)$ with dimensions of $(\text{mass})^D$, which depends on $n$ massive variables $q_1, \ldots, q_n$. Let us pick one of these variables — say $q_1$, though it could be any $q_i$ or any combination with the dimensions of mass — and rename it "$Q$," and then rewrite the others as dimensionless ratios $\theta_2 \equiv q_2/Q$, etc.

If the usual (unique) DA applied, $\sigma$ would have to take the form

$$\sigma(Q, \theta_2, \ldots, \theta_n) = Q^D S\left(\theta_2, \ldots, \theta_n\right), \tag{2.13}$$

where the function $S$ is dimensionless and depends only on the $(n-1)$ ratios. This form will no longer hold because of the one-parameter ambiguity of the theory, now embodied in $\mathcal{A}(\mu)$. Reapplying DA with this in mind the correct form must be

$$\sigma(Q, \theta_2, \ldots, \theta_n) = Q^D S\left(\theta_2, \ldots, \theta_n; \frac{Q}{\mu}, \mathcal{A}(\mu)\right), \tag{2.14}$$

where $S$ must be independent of $\mu$, the arbitrary massive parameter which serves as the argument of $\mathcal{A}$.

Let us analyze the form of Eq. (2.14). Firstly, since $\mu$ is arbitrary, we are free to pick $\mu = Q$ if we please. Hence, the dependence of $S$ on $Q/\mu$ and $\mathcal{A}(\mu)$ is equivalent to dependence on the single variable $\mathcal{A}(Q)$.

> **One can** also show this in a more laborious way. Equation (2.14) is similar in many respects to Eq. (2.2), but without the latter's recursive nature, so one can repeat the analysis in Eqs. (2.7)–(2.12) to show that $S$ depends on $Q/\mu$ and $\mathcal{A}(\mu)$ only through the combination $K(\mathcal{A}(\mu)) + \ln(Q/\mu)$. This variable, by Eq. (2.5), is just $K(\mathcal{A}(Q))$.

Another way of expressing the $\mu$-independence of Eq. (2.14) is the following. Define a function $\tilde{S}$ of $(n+1)$ independent variables

$$\tilde{S} = \tilde{S}\left(\theta_2, \ldots, \theta_n; \frac{Q}{\mu}, A\right), \tag{2.15}$$

which reduces to $S$ when the extra variable $A$ is set equal to $\mathcal{A}(\mu)$. $\tilde{S}$ becomes independent of $\mu$ only when $A = \mathcal{A}(\mu)$. Translated into mathematics, that statement becomes

$$\left.\frac{d\tilde{S}}{d\mu}\right|_{\text{at } A=\mathcal{A}(\mu)} = \left.\left(\frac{\partial \tilde{S}}{\partial \mu} + \frac{\partial \tilde{S}}{\partial A}\frac{dA}{d\mu}\right)\right|_{\text{at } A=\mathcal{A}(\mu)} = 0. \tag{2.16}$$

Multiplying by $\mu$ and remembering that $\mathcal{A}(\mu)$ is specified by

$$\mu\frac{d\mathcal{A}}{d\mu} = \mathcal{B}(\mathcal{A}), \tag{2.17}$$

Eq. (2.16) becomes

$$\left(\mu\frac{\partial}{\partial\mu} + \mathcal{B}(\mathcal{A})\frac{\partial}{\partial A}\right)\tilde{S}\Bigg|_{\text{at } A=\mathcal{A}(\mu)} = 0. \qquad (2.18)$$

The reader will probably recognize this as a renormalization group (RG) equation of some kind. Indeed, the equation can be applied to $\sigma$ itself:

$$\left(\mu\frac{\partial}{\partial\mu}\Bigg|_{\mathcal{A}} + \mathcal{B}(\mathcal{A})\frac{\partial}{\partial\mathcal{A}}\right)\sigma = 0. \qquad (2.19)$$

Here the notation is perhaps a bit sloppy and may need a little explanation. In Eq. (2.18), it was evident that $\mu$ and $A$ were independent variables, so that each partial derivative is taken with the other variable held constant. Then, afterwards, one is to replace $A$ with $\mathcal{A}(\mu)$. In Eq. (2.19) we dispense introducing $A$, so we have added "$\ldots|_{\mathcal{A}}$"as a reminder that the $\mu$ partial derivative is to be taken pretending that $\mathcal{A}$ is a constant, despite the fact that $\mathcal{A}$ is a function of $\mu$.

There is nothing magic or mysterious about this renormalization-group (RG) equation. Solving this partial differential equation only leads back to Eq. (2.14), together with the verbal statement that $\sigma$ is independent of $\mu$. The form (2.14) followed simply from Dimensional Analysis, the only subtlety being the theory's one-parameter ambiguity.

That subtlety requires a distinction between *canonical* dimensions and *scale* dimensions. The canonical dimension of $\sigma$ is just $D$, and it counts both the $Q$ and $\mu$-dependences since both are *massive* parameters. The *scale* dimension, however, only counts the dependence of $\sigma$ on the physical scale $Q$:

$$\mathcal{D}_{(\sigma)} \equiv \frac{Q}{\sigma}\frac{d\sigma}{dQ}, \qquad (2.20)$$

Since $S \equiv \sigma/Q^D$ is a dimensionless function, we know from DA that it can depend on $Q$ only through the ratio $Q/\mu$, as in Eq. (2.14), so that

$$\frac{Q}{S}\frac{dS}{dQ} = -\left.\frac{\mu}{S}\frac{dS}{d\mu}\right|_{\mathcal{A}}. \tag{2.21}$$

Note that we must hold $\mathcal{A}(\mu)$ fixed here. Hence, $\mathcal{D}_{(\sigma)}$ can be expressed as $D - \left.\frac{\mu}{\sigma}\frac{\partial\sigma}{\partial\mu}\right|_{\mathcal{A}}$. Using the RG equation (2.19), we then find

$$\mathcal{D}_{(\sigma)} = D + \frac{\mathcal{B}(\mathcal{A})}{\sigma}\frac{\partial\sigma}{\partial\mathcal{A}}. \tag{2.22}$$

## 2.6. Summary

We began by asking our theory, considered as a Black Box, about a specific, dimensionless physical quantity $\mathcal{A}(Q)$. Naïve DA, which would force the answer to be a constant, does not apply because the theory's answer is not unique. Rather, "one-parameter ambiguous DA" applies, requiring the answer to be of the form of Eq. (2.1):

$$Q\frac{d\mathcal{A}}{dQ} = \mathcal{B}(\mathcal{A}), \tag{2.23}$$

where the function $\mathcal{B}(\mathcal{A})$ is a dimensionless function of a dimensionless argument, and is fully specified by the theory. The one-parameter ambiguity arises from the missing boundary condition — which allows a massive constant $\Lambda$ to creep in. When asked about other physical quantities, the Black Box specifies them in terms of the first quantity $\mathcal{A}(\mu)$ at some arbitrary mass scale $\mu$. The fact that the new physical quantity, $\sigma$, is independent of the arbitrary mass scale $\mu$ is expressed by Eq. (2.19):

$$\left(\left.\mu\frac{\partial}{\partial\mu}\right|_{\mathcal{A}} + \mathcal{B}(\mathcal{A})\frac{\partial}{\partial\mathcal{A}}\right)\sigma = 0. \tag{2.24}$$

There is an obvious artificial asymmetry in the preceding description of the theory. One particular physical quantity $\mathcal{A}$ is singled out as "special" and all others are expressed in terms of it. This is an arbitrary choice: we could choose *any* physical quantity to play the role of the "special" one. If we had a Black Box that solved the theory exactly, there would be no problem here: whatever description of the theory we adopted, the results would be entirely equivalent. In fact, the special quantity does not need to be a *physical* quantity. It could be defined in theoretical, rather than experimental terms. A natural choice, especially in the context of perturbation theory, is to use some kind of "renormalized coupling constant." If the theory could be solved exactly, it would not matter how we chose to define it. At finite orders of perturbation theory, however, there is a problem. Before we return to that issue — the main topic of this book — let us digress to consider unphysical quantities, such as Green's functions.[a]

## 2.7. Unphysical Quantities and "Anomalous Dimensions"

Many of the quantities we deal with in field theories are not, even in principle, directly measurable quantities. These useful, but unobservable, quantities, such as Green's functions or proper vertices, often suffer from bad diseases. For instance, they can be infrared divergent, and they can depend on unphysical parameters. In gauge theories they can depend on the gauge-fixing parameter. (That is an additional complication, ignored here; it is discussed in Appendix 2.B.) The relevant point here is that such unphysical quantities can depend on $\mu$.

Of course, the advantage of having the Black Box is that we do not need to trouble ourselves with unphysical quantities. Such quantities are only needed as intermediate steps in calculating physical quantities — which our Black Box now enables us to obtain directly, with no labour. However, out of curiosity, we might like to ask the Black Box to calculate an unphysical quantity for us. The answer, though, will depend, not only on the variables we specify, but also on unphysical variables, including $\mu$. The "meaning" of this variable is buried in the Manufacturer's Specification for the Black Box which describes the technical details of the renormalization prescription, gauge choice, etc., that the machine uses in its calculations. We need not be concerned with these details here.

---

[a]The next section is useful background for Part III, but is not needed for Part II.

To begin with, let us consider a hypothetical example in which a specific unobservable quantity $\Gamma(q_1, \ldots, q_n; \mu)$, with dimensions of $(\text{mass})^D$, happens to depend on $\mu$ in a particularly simple way, namely

$$\Gamma(q_i; \mu) = \mu^\gamma \, G(q_i; \mu, \mathcal{A}(\mu)), \tag{2.25}$$

where the exponent $\gamma$ is a numerical constant and $G$ is independent of $\mu$.

The statement that $\Gamma$ has the form of Eq. (2.25) can be translated into a RG equation. As in the previous section let us regard $\Gamma$ as the special case of $\tilde{\Gamma}(q_i; \mu, A)$ when $A = \mathcal{A}(\mu)$, so that, as an identity,

$$\frac{d\Gamma}{d\mu} = \left[ \frac{\partial\tilde{\Gamma}}{\partial\mu}\bigg|_A + \frac{dA}{d\mu}\frac{\partial\tilde{\Gamma}}{\partial A}\bigg|_\mu \right]\bigg|_{\text{at } A=\mathcal{A}(\mu)}. \tag{2.26}$$

But, from Eq. (2.25),

$$\mu\frac{d\Gamma}{d\mu} = \gamma\,\Gamma. \tag{2.27}$$

Substituting this in Eq. (2.26), and invoking Eq. (2.17), gives

$$\left[ \mu\frac{\partial}{\partial\mu} + \mathcal{B}(A)\frac{\partial}{\partial A} - \gamma \right]\tilde{\Gamma}\bigg|_{\text{at } A=\mathcal{A}(\mu)} = 0, \tag{2.28}$$

which is recognizable as a typical RG equation, containing an "anomalous dimension" term, $\gamma$.

Again, there is no particular magic in having a partial differential equation. The content of this equation is much better expressed by its solution, which is Eq. (2.25) together with the verbal statement that $G$ is independent of $\mu$. However, the virtue of Eq. (2.28) is that this form generalizes to apply to $\Gamma$'s which depend arbitrarily on $\mu$. For this general case $\gamma$ must now be regarded as a function *defined* by Eq. (2.27).

From DA alone there is no reason why $\gamma$ cannot depend on $\frac{q_1}{\mu}, \frac{q_2}{q_1}, \ldots, \frac{q_n}{q_1}$ as well as $\mathcal{A}(\mu)$. If this happened in practice, then the RG equation would hardly be very useful. Fortunately, in field theory the $\gamma$'s come from the renormalization constants (the $Z$'s), so they have no dependence on any of the physical variables $q_1, \ldots, q_n$, and hence, by DA, they can only depend on $\mu$ through $\mathcal{A}(\mu)$; thus, $\gamma = \gamma(\mathcal{A}(\mu))$. This happy circumstance is not altogether surprising: the various unmeasurable quantities must be able to combine together to produce observable quantities in such a way that their $\mu$ dependence mutually cancels. In consequence, the $\mu$-dependence of each quantity has to be relatively simple; e.g., in a Green's function the $\mu$-dependence must factor into separate $\mu$-dependences for each external line.

**Equation (2.28)** is derived field-theoretically by considering the renormalization of the quantity $\Gamma$. The essential ingredients of the derivation are (i) the $\mu$-independence of the bare quantity, (ii) the cutoff independence of the renormalized quantity, (iii) their "proportionality" through some infinite renormalization constant ($\Gamma_{\text{ren}} = Z\Gamma_{\text{bare}}$) and (iv) Dimensional Analysis.

The solution of Eq. (2.28) when $\gamma$ is a function of $\mu$, but not of $q_1, \ldots, q_n$, is

$$\Gamma(q_i; \mu) \equiv \tilde{\Gamma}(q_i; \mu, \mathcal{A}(\mu)) = \mu^{[\gamma]} \, G(q_i; \mu, \mathcal{A}(\mu)), \tag{2.29}$$

where $G$ is independent of $\mu$, and the notation $\mu^{[\gamma]}$ means

$$\mu^{[\gamma]} \equiv \exp \int^{\mu} \frac{d\mu'}{\mu'} \, \gamma(\mathcal{A}(\mu')). \tag{2.30}$$

It is easily verified that the $\mu$-dependence of $\Gamma$ satisfies Eq. (2.27), the defining equation for $\gamma$. Using Eq. (2.23), but with $\mu$ as the argument of $\mathcal{A}$, we may express $\mu^{[\gamma]}$ as

$$\mu^{[\gamma]} = \exp \int^{\mathcal{A}(\mu)} dx \, \frac{\gamma(x)}{\mathcal{B}(x)}, \tag{2.31}$$

where $x$ is a dummy "$\mathcal{A}$" argument.

The scale dimension of $\Gamma$,

$$\mathcal{D}_{(\Gamma)} \equiv \frac{Q}{\Gamma} \frac{d\Gamma}{dQ}, \tag{2.32}$$

is (unlike $\gamma$) a physical quantity. That is because the wavefunction-renormalization constant $Z$, multiplicatively renormalizing $\Gamma$, will cancel out since it is independent of the momentum arguments $q_i$, and hence $Q$ independent. Generalizing the discussion leading to Eq. (2.22) one finds

$$\mathcal{D}_{(\Gamma)} = D - \gamma(\mathcal{A}) + \frac{\mathcal{B}(\mathcal{A})}{\Gamma} \frac{\partial \Gamma}{\partial \mathcal{A}}, \qquad (2.33)$$

where $\mathcal{A} = \mathcal{A}(Q)$ here. The $-\gamma(\mathcal{A})$ term arises from the $\mu^{[\gamma]}$ factor in $\Gamma$.

*To summarize:* the non-trivial content of the RG equation, Eq. (2.28), is that the anomalous dimension $\gamma$ depends on $\mu$ only, and only via $\mathcal{A}(\mu)$. This implies that the $\mu$-dependence of the unphysical quantity $\Gamma(q_i; \mu)$ *factorizes* and is not intertwined with the $q_i$-dependence.

## 2.8. Illustration: QCD in Leading Order

In this section, we consider leading order in perturbation theory. For the present pedagogical purposes, we will ignore the fact that leading order in perturbation theory is not an exact solution of the theory. (Alternatively, we can view it as effectively exact if we limit our ambition to predicting only the *leading* asymptotic behaviour in the $Q \to \infty$ limit.) Our aim here is to make contact with the usual textbook treatments of some classic QCD results.

As the reader may well have already observed, the formulas in the preceding sections look much more familiar if $\mathcal{A}(Q)$ is identified with $\alpha_s(Q)$, the QCD running coupling constant at an energy scale $Q$. This identification will indeed be made in this section. It does not matter here that $\alpha_s(Q)$ is not a proper observable, in that it is not defined in experimental terms. Its precise definition will not matter for the current discussion.

With $\mathcal{A}(Q) = \alpha(Q)$ (we omit the $s$ subscript henceforth) Eq. (2.1) is immediately recognizable as the relation between the coupling constant and the "$\beta$ function"

$$Q \frac{d\alpha}{dQ}(Q) = \beta(\alpha(Q)). \qquad (2.34)$$

**In the literature** there are several variants of this definition of $\beta$: sometimes it is defined in terms of $g(Q)$, where $\alpha = g^2/(4\pi)$, and sometimes the variable is $Q^2$ rather than $Q$. We use the above form in this section because it is perhaps the most common in textbooks. Later, however, we will adopt a slightly different definition of $\beta$ in terms of $a \equiv \alpha/\pi$.

The importance of the $\beta$ function is that it is the theory's oblique way of specifying $\alpha(Q)$. Notice that the sign of $\beta$ governs whether $\alpha(Q)$ increases or decreases with $Q$. In QCD, the $\beta$ function has a perturbation expansion which begins

$$\beta(\alpha) = -\beta_0 \alpha^2 + \cdots, \tag{2.35}$$

where $\beta_0$ is a positive number, calculable from Feynman diagrams. For a theory with $n_f$ flavours of (massless) quarks, $\beta_0 = (33 - 2n_f)/(6\pi)$. The $K$ function of Eqs. (2.3), (2.11) is then, to leading order,

$$K(\alpha) \equiv \int_\infty^\alpha \frac{d\alpha'}{-\beta_0 \alpha'^2} = \frac{1}{\beta_0 \alpha}. \tag{2.36}$$

Equations (2.5), (2.6) then yield the familiar results

$$\alpha(Q) = \frac{\alpha(\mu)}{1 + \beta_0 \alpha(\mu) \ln(Q/\mu)} \quad \text{for } any \ \mu, \tag{2.37}$$

$$\alpha(Q) = \frac{1}{\beta_0 \ln(Q/\Lambda)} \quad \text{for } some \ \Lambda. \tag{2.38}$$

As is clear from the earlier discussion, Eq. (2.38) is equivalent to Eq. (2.37) and not some kind of approximation to it. The decrease of $\alpha(Q)$ as $Q \to \infty$, due to $\beta(\alpha)$ being negative, is the justly famous "asymptotic freedom" property.

With $\mathcal{A}(Q)$ and $\mathcal{B}(\mathcal{A})$ replaced by $\alpha(Q)$ and $\beta(\alpha)$, respectively, Eq. (2.19) becomes precisely the conventional RG equation for a physical quantity;

$$\left( \mu \frac{\partial}{\partial \mu} \bigg|_\alpha + \beta(\alpha) \frac{\partial}{\partial \alpha} \right) \sigma = 0. \tag{2.39}$$

As remarked before, this equation means that the physical quantity $\sigma$ is actually independent of $\mu$; the explicit $\mu$ dependence cancels with the implicit $\mu$ dependence via $\alpha(\mu)$. From one-parameter-ambiguous DA we could have written down immediately the solution of this partial differential equation as Eq. (2.14).

For example, consider the physical quantity $R_{e^+e^-}(Q) \equiv \sigma_{\text{tot}}(e^+e^- \rightarrow \text{hadrons})/\sigma(e^+e^- \rightarrow \mu^+\mu^-)$ at total center-of-mass energy $Q$. According to one-parameter-ambiguous DA, this dimensionless function of one massive variable $Q$ must have the form

$$R_{e^+e^-}(Q) = f\left(\frac{Q}{\mu}, \alpha(\mu)\right) \quad \text{for any } \mu. \tag{2.40}$$

$R_{e^+e^-}(Q)$ has a sensible free-field-theory limit, namely the parton-model result $\left(3\sum_i q_i^2\right)$, where $q_i$ is the electric charge of the $i$th quark. (Note that the parton-model result is a $Q$-independent constant because ordinary DA then applies.) If we write

$$R_{e^+e^-}(Q) \equiv \left(3\sum_i q_i^2\right)(1+\mathcal{R}), \tag{2.41}$$

then $\mathcal{R}$ represents the QCD correction. It must be a function only of $\frac{Q}{\mu}$ and $\alpha(\mu)$, and has a perturbation expansion:

$$\mathcal{R} = r_0\left(\frac{\alpha(\mu)}{\pi}\right)\left(1 + r_1\left(\tfrac{Q}{\mu}\right)\left(\frac{\alpha(\mu)}{\pi}\right) + \cdots\right) \tag{2.42}$$

with coefficients that are calculable from Feynman diagrams ($r_0$ turns out to be 1). The perturbative coefficients, $r_1, \ldots$, depend on the ratio $\frac{Q}{\mu}$. This dependence, as we shall see in the next chapter, is through $\ln\frac{Q}{\mu}$ terms, so that if $Q$ becomes very large relative to $\mu$, the perturbative coefficients become horribly large. Thus, in practice — not having an exact result, but only the first few terms of a perturbation series — our choice of the arbitrary scale $\mu$ actually matters. Rather than fixing $\mu$ once-and-for-all, we should allow it to "run" with $Q$, so that $R_{e^+e^-}(Q)$ depends only on $\alpha(Q)$:

$$\mathcal{R} = \left(\frac{\alpha(Q)}{\pi}\right)\left(1 + r_1(\mu=Q)\left(\frac{\alpha(Q)}{\pi}\right) + \cdots\right) \tag{2.43}$$

This is known as "renormalization-group-improved" perturbation theory. In an asymptotically free theory the "running coupling constant" $\alpha(Q)$ tends to zero as $Q$ tends to infinity, so that RG-improved perturbation theory is good at high energies. Thus, we can definitely say that the leading asymptotic behaviour is $\mathcal{R} \sim \alpha(Q)/\pi \sim 1/(\pi\beta_0 \ln Q)$.

However, the RG/DA argument only sets the *scale* for the running coupling constant. One can equally well show, by putting $\mu = nQ$ in Eq. (2.40), that $R_{e^+e^-}(Q)$ is a function only of $\alpha(nQ)$, where $n$ could be any constant (or, indeed, any function of $\alpha(\mu)$). If we had exact results, or if we only asked about $Q \to \infty$, the choice of $n$ would not matter at all. But in practice, since we want to get quantitative results from a truncated perturbation series, the choice of $n$ *does* matter. This choice is entangled with the other RS choices involved in the definition of $\alpha$, and is just one aspect of the wider RS-dependence problem.

> **At leading order,** when the correction coefficient $r_1$ is unknown, one can only guess at a "reasonable" $n$: Leading order is thus only a *qualitative* approximation. While good phenomenology can be done — and was done — with just leading-order results, *quantitative* results require at least next-to-leading order calculations.

## Appendix 2.A: Theories with Masses

Our discussion in this chapter was specific to massless theories. It might at first seem that, if the Lagrangian contains a mass term, then all the previous arguments are undermined. However, this is not true since the bare mass, being ill-defined (infinite or infinitesimal depending on regularization), cannot directly provide a finite mass scale, any more than the bare coupling constant could.

In fact, the previous picture still holds, except that the theory now has a *two*-parameter ambiguity. Generalizing the previous procedure, one can isolate these ambiguities in two renormalized quantities. These are most naturally chosen to be a coupling constant $\alpha(\mu)$ and a "running mass" $m(\mu)$. Each of these functions is specified via a first-order differential equation, whose boundary condition must be determined by experiment — through measurements of $\Lambda$ and the physical mass $m_{\text{phys}}$, respectively.

Other quantities must then be expressed in terms of $\alpha(\mu)$, $m(\mu)$, and $\mu$, leading to generalizations of the RG equations (2.19) and (2.28) containing an extra $-\gamma_m m(\partial/\partial m)$ term, where $\gamma_m$ is somewhat analogous to the $\beta$ function: $\mu\, dm/d\mu = -\gamma_m m(\mu)$. This leads naturally to Weinberg's form of the RG equation. A proper field-theoretic derivation of this equation gives the vital extra information that both $\gamma_m$ and the $\beta$ function depend only on $\alpha(\mu)$.

To derive the Weinberg RG equation field-theoretically, one works in a "mass-independent" renormalization scheme (e.g., the $\overline{\text{MS}}$ scheme), which does not involve the physical mass in its definition. Alternatively, one may fix the mass to be the physical mass and use the original formulation due to Callan and Symanzik, but this involves the specifically field-theoretic concept of mass insertions in Green's functions.

Although the focus in this book is mostly on massless theories, the optimization procedure of Part II can be applied directly to a massive theory if the mass renormalization uses the physical mass. However, there might be further advantages in generalizing the procedure so as to optimize both the running coupling and the running mass.

## Appendix 2.B: Gauge Dependence in Gauge Theories

Here we return to the problem of the gauge dependence of unphysical quantities which was avoided in Sec. 2.7. This can be treated by analogy with Appendix 2.A's discussion of mass terms. To quantize a gauge theory one must add a gauge-fixing term to the Lagrangian. This step effectively introduces another sort of bare coupling constant — the bare gauge parameter $\xi$ — which affects only unphysical quantities. The bare gauge parameter must itself be renormalized, leading to a renormalized gauge parameter $\xi(\mu)$. The RG equation for unphysical quantities will contain, in addition to an anomalous dimension, a term $\delta(\alpha, \xi)(\partial/\partial\xi)$, where $\delta$ is the analogue of the $\beta$ function: $\mu\partial\xi/\partial\mu \equiv \delta(\alpha, \xi)$, with the partial derivative taken at constant $\alpha$. Observable quantities, being gauge invariant, do not depend on $\xi(\mu)$, so their RG equations do not need this term.

Some RS's explicitly involve the gauge choice in their definition. In such schemes, there are two ways in which one might define the $\beta$ function, depending on whether the derivative in $\mu \partial \alpha / \partial \mu$ is taken holding constant the *bare* or the *renormalized* gauge parameter. It is important for our later analysis, which will use only renormalized quantities, that the latter definition is used.

## Chapter 3

# Renormalization as Reparametrization

## 3.1. Introduction

This chapter aims to describe renormalization conceptually, with technicalities kept to a minimum. It is intended to complement the presentations in QFT textbooks, which should be consulted for more concrete, technical details.

Consider a physical quantity in QFT, such as a scattering cross section $\sigma$, that can be calculated in perturbation theory. It will be a function, not only of various kinematic variables (the centre-of-mass energy, the scattering angle, etc.), but also of the parameters in the theory's Lagrangian, such as the particle masses and coupling constants. For simplicity, let us consider a theory with just one mass parameter $m$ and one coupling constant $g$.

The physical quantity $\sigma$ can be calculated in perturbation theory from the relevant Feynman diagrams. The leading order result comes from one or more "tree" diagrams, and is seemingly a well-defined, finite result, $\sigma_0$. However, the higher-order terms come from diagrams involving loops. These loops give rise to integrations over the loop's 4-momentum and these integrals often fail to converge in the ultraviolet. Thus, at first sight, the result that emerges from perturbation theory seems meaningless:

$$\sigma = \sigma_0 + g^2(\infty?!?) + \cdots . \tag{3.1}$$

Historically, in quantum electrodynamics (QED), the leading term $\sigma_0$ was found to agree quite nicely with experiment, but when theorists calculated the "correction term" and found that it was infinite there was consternation.

The problem lies entirely in the fact that we are trying to parametrize $\sigma$ in terms of the mass and coupling-constant parameters present in the Lagrangian — and these are *not* the mass and coupling strength that one could measure in any actual experiment. We will henceforth add a "B" subscript to stress that these are "bare" quantities. They are not measurable quantities at all, and therefore need not be finite. There is no need to be frightened of this. It is one of the great lessons of modern physics that only the *physically observable* quantities of a theory need be real and finite (and agree with experiment); successful theories may well involve other, unmeasurable objects that are complex, or infinite, infinitesimal, ambiguous, or otherwise bizarre. For example, quantum mechanics — unlike classical mechanics — intrinsically involves *complex*-valued amplitudes, but all its predictions for observable quantities are *real*. In quantum field theory, there turn out to be various theoretical objects that are not even finite, but that need not bother us provided all the physical predictions are finite, real, and sensible (and agree with experiment).

The bare mass $m_B$ is the mass that the particle would have in the unperturbed theory — that is, in the free-field theory obtained by setting $g_B = 0$. In the full, interacting theory, however, the propagator is given by a sum over all possible two-legged Feynman diagrams – representing the particle interacting with itself and with fluctuations of the vacuum (see Fig. 3.1). These interactions shift the position of the propagator's pole, and hence change the particle's mass. The particle's physical mass $m_{\text{phys}}$ — the mass measurable in an experiment — corresponds to the position of the pole in this full propagator.

The "bare" coupling constant $g_B$ is the coupling constant for a "free" particle, or rather, a particle that is allowed to interact once, and once only. That is a somewhat absurd and unphysical concept, because as soon as we "turn on" the perturbation the particle will interact, with itself and with vacuum fluctuations, arbitrarily many

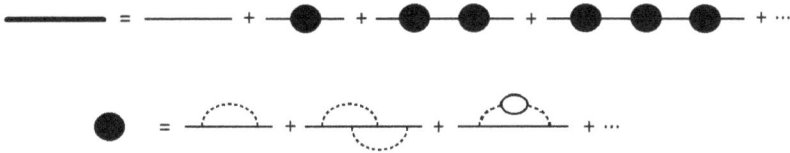

Fig. 3.1. The full propagator is a geometric sum of self-energy insertions, where the self-energy is the sum of all 1-particle irreducible diagrams — ones that cannot be separated into two pieces by cutting just a single line. (The external legs of the latter diagrams are "amputated"; they have no propagator factors associated with them.) For illustrative purposes we show diagrams in a QED-like theory.

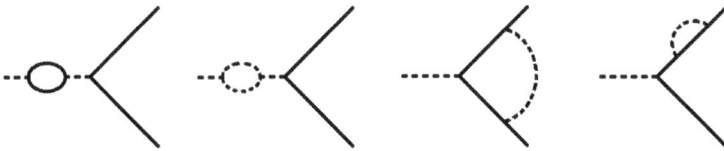

Fig. 3.2. The effective coupling strength will be affected by interactions that "dress" the bare vertex. A few illustrative examples are shown.

times. Thus, any attempt to measure an "effective coupling strength" $g_{\text{eff}}$ (however one might choose to define it, experimentally) will be measuring something that involves a sum of infinitely many Feynman diagrams (see Fig. 3.2).

For the theory to be viable, the *measurable* quantities $m_{\text{phys}}$ and $g_{\text{eff}}$ need to be finite. But there is no reason to demand that $m_B$ and $g_B$ be finite — because there is no way, even in principle, to measure them. The problem of infinities in the calculation of a physical quantity $\sigma$ is simply an artefact of the result being parametrized in terms of $m_B$ and $g_B$, which are not themselves finite quantities. All we need to do is to *reparametrize* the result for $\sigma$ in terms of parameters that *are* finite, such as $m_{\text{phys}}$ and $g_{\text{eff}}$. In fact, it is not necessary to use parameters that are truly measurable quantities; it is enough that we use suitable "renormalized" parameters $m_R, g_R$ that are defined so that they are related in a finite and definite way to the physically measurable $m_{\text{phys}}$ and $g_{\text{eff}}$.

**RENORMALIZATION = REPARAMETRIZATION**

$$
\begin{array}{c}
\text{Theory} \\
\mathcal{L}(m_B, g_B)
\end{array}
\longrightarrow
\left\{
\begin{array}{c}
\sigma_1(m_B, g_B) \\
\cdot \\
\sigma_N(m_B, g_B)
\end{array}
\right\}
\begin{array}{c}
m_R = f_m(m_B, g_B) \\
\text{-----} \longrightarrow \\
g_R = f_g(m_B, g_B)
\end{array}
\left\{
\begin{array}{c}
\tilde{\sigma}_1(m_R, g_R) \\
\cdot \\
\tilde{\sigma}_N(m_R, g_R)
\end{array}
\right\}
\longrightarrow \text{Experiment}
$$

*Predictions for N physical quantities $\sigma_1, \ldots, \sigma_N$ are calculated from the theory. Initially, when parametrized by $m_B, g_B$, they contain divergences. However, once reparametrized in terms of suitably defined "renormalized" parameters $m_R, g_R$, they become finite functions of finite parameters. They can then be fitted to experimental data, providing $N - 2$ tests of the theory, together with "best-fit" values for the $m_R, g_R$ parameters.*

If we had a Black Box that could solve the theory exactly, then this situation would be evident. The physical predictions of the theory are, in fact, finite, but to make this fact manifest one must make a substitution, eliminating $m_B, g_B$ in favour of $m_R, g_R$. It is just a substitution and nothing is being thrown away or "swept under the rug." The term "renormalization" is really a misnomer, and "reparametrization" would be a better name.

**The above statement** should be qualified by saying that it applies to the *physical* quantities of the theory. Green's functions are not made finite by reparametrization alone; they also need to be rescaled by a suitably infinite overall factor. That is, they need an actual change of normalization, a true renormalization. Each leg of an $n$-point Green's function needs a suitably infinite factor $Z_{\mathrm{wf}}^{-1/2}$, where $Z_{\mathrm{wf}}$ is the so-called "wavefunction renormalization" constant, so that the whole $n$-point function is rescaled by $Z_{\mathrm{wf}}^{-n/2}$. (See Appendix 3.A.)

If we could solve the theory exactly, it would not matter exactly how we define the $m_R, g_R$ parameters. Any definition that does the job of manifesting the finiteness of the theory's predictions would be acceptable. Changing from one definition to another would change the appearance of our results, but not their content. However, when we use perturbation theory, our $g_R$ parameter plays a double role: It is not only one of the free parameters of the theory (taking over that role from $g_B$) it is also the expansion parameter of our approximation procedure. Because of this second role, the choice of definition of $g_R$ does matter in practice. We return to this central concern in a moment.

## 3.2. Regularization

We have talked quite cavalierly about "divergences" and "infinities," but in order to be doing proper mathematics we need a procedure known as "regularization." This step involves many technicalities — which often dominate textbook accounts of renormalization and tend to obscure what is really going on.

A "regularization" is some systematic, consistent, mathematical modification of the theory that renders the divergent integrals convergent. One example is some kind of "ultraviolet cutoff" procedure that puts an upper limit on the magnitude of the momentum allowed in the loop integrals. We will use "$M_{\mathrm{uv}}$" to denote the ultraviolet cutoff parameter. Other regularization methods modify the free propagator, using a large parameter $M_{\mathrm{uv}}$ that acts much like a cutoff. Another type of regularization is dimensional regularization, which formulates the theory in $d$ space–time dimensions, where $d$ is allowed to be non-integer and is taken to be $4 - \epsilon$, with $\epsilon \to 0$. We will return to discuss dimensional regularization later.

Regularization is like scaffolding that holds up a rickety building while its structure is being strengthened. It replaces meaningless divergent integrals with well-defined mathematical expressions that are finite, but which would become arbitrarily large in the limit $M_{\mathrm{uv}} \to \infty$. After one has made the reparametrization step — eliminating $m_B, g_B$ in favour of $m_R, g_R$ — one finds that cancellations occur, so that the limit $M_{\mathrm{uv}} \to \infty$ can then be taken, yielding manifestly finite results.

It is not necessary that the regularized theory be a physically acceptable theory. Indeed, the regularized theory is often very ugly and may violate some symmetries of the original theory. One has to take care that important symmetries, such as Lorentz invariance and gauge invariance, are indeed recovered when the $M_{\mathrm{uv}} \to \infty$ limit is taken.

**However,** there are some cases of "**quantum anomalies**," where a symmetry of the bare Lagrangian — which would certainly be a symmetry of the corresponding classical field theory — cannot be maintained in the regularized, quantum theory even in the $M_{uv} \to \infty$ limit. Dimensional transmutation, whereby a scale-invariant Lagrangian gives rise to a theory with a characteristic scale $\Lambda$, is one example of a quantum anomaly. Chiral symmetry is another case where anomalies arise.

It is very important to be clear that regularization, though a technical necessity, is not *physics* — just as scaffolding is not architecture. In the end the regularization is removed, revealing the theory's finite predictions — just as scaffolding is eventually taken down to reveal the newly renovated building.

## 3.3. Defining a Renormalized Coupling Constant

There are many, many ways of going about defining a renormalized coupling constant $g_R$. One traditional approach, known generically as "momentum subtraction" (or MOM) schemes, starts from the observation that, since the bare coupling constant $g_B$ corresponds to the bare 3-particle vertex, it is natural to define $g_R$ in terms of a "dressed" 3-particle vertex, involving a sum over all possible Feynman diagrams with three (off-shell) legs. Conceptually, this idea is helpful — but attempting to develop it into a specific definition of $g_R$ leads one into a deep quagmire of technicalities. For example, in QCD, the bare coupling $g_B$ is associated with various vertices — quark–quark–gluon, 3-gluon, and ghost–ghost–gluon vertices, as well as the 4-gluon vertex, proportional to $g_B^2$. Which of these vertices, or combination thereof, should we choose? Then there is the question of the momentum carried by each leg: Unlike the bare vertex, the dressed vertex depends non-trivially on the momentum configuration. There are many other technicalities, such as the spin-index decomposition, etc. As an example, a typical MOM scheme might define $g_R$ as "the coefficient of the Dirac matrix $\gamma^\mu$ in the decomposition ... of the quark–quark–gluon vertex at the "symmetric point" where the 4-momenta $p_1^\mu, p_2^\mu, p_3^\mu$ satisfy

$$p_1^2 = p_2^2 = p_3^2 = -\tfrac{3}{4}\mu^2, \quad p_1.p_2 = p_2.p_3 = p_3.p_1 = \tfrac{1}{4}\mu^2, \qquad (3.2)$$

in the Feynman gauge with ...." (We have suppressed some of the more horrific technical details with "...".) Note that the renormalization conditions can be specified without reference to any particular regularization procedure.

The most important aspect of this, or any such definition, is that $g_R$ will depend on a scale $\mu$, known as the "renormalization point" or "renormalization scale," that sets the overall energy-scale of the dressed vertex. That is

$$g_R = g_R(\mu, \text{ other choices}). \tag{3.3}$$

We will later use the terminology these other choices are the "renormalization prescription" (RP). When we say that $g_R$ depends on renormalization scheme (RS), we mean that it depends on $\mu$ and the "other choices" made in defining $g_R$. Indeed, for physical quantities, "specifying a RS" means the same thing as "defining a renormalized coupling constant, $g_R$."

> **One should beware** of claims to have defined an "RS-invariant renormalized coupling constant." That is simply a contradiction in terms. The definition of a coupling constant is a renormalization scheme, and vice versa. We return to this point at the end of the chapter.

Let us next consider calculating $g_R$ in perturbation theory, assuming that we have adopted some specific RP. The first term is just $g_B$, but there are then corrections involving Feynman diagrams with two more vertices (and hence two more powers of $g_B$) and all contain one loop. These loops result in integrals that are typically divergent and so will need regularization. That is, we will find a result of the form

$$g_R = g_B(1 + g_B^2 \tilde{I}(\mu) + \cdots), \tag{3.4}$$

where $\tilde{I}(\mu)$ is some divergent integral, or sum of divergent integrals, rendered large but finite by the regularization. $\tilde{I}(\mu)$ will depend on the specific choices in the $g_R$ definition adopted; in particular it will depend on the scale $\mu$. It is convenient to work with the square of $g$ and absorb some $\pi$ factors by defining

$$a \equiv \frac{\alpha_s}{\pi} \equiv \frac{g^2}{4\pi^2} \tag{3.5}$$

for both $B$ and $R$ cases. Squaring Eq. (3.4) then leads to

$$a_R = a_B(1 + a_B I(\mu) + \cdots), \qquad (3.6)$$

where $I(\mu) = 8\pi^2 \tilde{I}(\mu)$. Inverting this equation ("reversion of a power series") gives

$$a_B = a_R(1 - a_R I(\mu) + \cdots). \qquad (3.7)$$

## 3.4. Renormalizing a Physical Quantity

Now consider the calculation of some physical quantity, such as a two-particle scattering cross section $\sigma$. It will depend on some kinematic variables, which we can take to be the total centre-of-mass energy $Q$ and scattering angle $\theta$. At lowest order the result is given by a tree diagram, but then there are corrections from one-loop diagrams, which again will have loop integrations that in general do not converge, in the absence of regularization. Thus, we will find a result of the form

$$\sigma = (\text{factor}) \left(g_B^2\right)^{\text{P}} (1 + g_B^2 \, \tilde{J}(Q, \theta) + \cdots), \qquad (3.8)$$

where $\tilde{J}$ is some sum of divergent integrals. (That is, $\tilde{J}$ depends sensitively on the regularization parameter $M_{\text{uv}}$, and goes to infinity as $M_{\text{uv}} \to \infty$.) For two-particle scattering, the power P would be 2, but more generally it could take other values and "$\theta$" could represent a set of several dimensionless variables. Converting from $g_B$ to $a_B \equiv g_B^2/(4\pi^2)$, and removing an overall factor, we can define a physical quantity $\mathcal{R}$ in the general form

$$\mathcal{R} = a_B^{\text{P}} \left(1 + a_B \, J(Q, \theta) + \cdots\right), \qquad (3.9)$$

with $J = 4\pi^2 \tilde{J}$.

Renormalizing this result is simply a matter of eliminating $a_B$ in favour of $a_R$ using Eq. (3.7). We spell out the simple algebra explicitly:

$$\mathcal{R} = (a_R(1 - a_R \, I(\mu) + \cdots))^{\text{P}} (1 + a_R \, J(Q, \theta) + \cdots)$$
$$= a_R^{\text{P}} (1 - \text{P} \, a_R I(\mu) + \cdots) (1 + a_R \, J(Q, \theta) + \cdots) \qquad (3.10)$$
$$= a_R^{\text{P}} (1 + a_R (J(Q, \theta) - \text{P} I(\mu)) + \cdots).$$

The "$+\cdots$" terms all have two or more factors of $a_R$ relative to the 1, so that formally they are "$O(a_R^2)$." What happens, in a renormalizable theory, is that the divergences in $J(Q,\theta)$ cancel with those of $\mathrm{P}I(\mu)$. Thus, we may now take the limit where the regularization goes away to obtain a finite coefficient

$$r_1 = r_1(Q,\theta,\mu) \equiv \lim_{M_{\mathrm{uv}}\to\infty} [J(Q,\theta) - \mathrm{P}I(\mu)]. \qquad (3.11)$$

Similar cancellations occur in higher orders, so that the result appears as a power series in $a \equiv a_R$ that has finite coefficients:

$$\mathcal{R} = a^{\mathrm{P}}(1 + r_1 a + \cdots). \qquad (3.12)$$

Here we have dropped the $R$ subscript, leaving it understood that $a$ stands for the (renormalized) couplant $g_R^2/(4\pi^2)$. We use the word "couplant" because the phrase "renormalized coupling constant" is both cumbersome and misleading — $a$ is not a constant; it is a function of the renormalization scale $\mu$, and of the "other choices" made in its definition, as indicated in Eq. (3.3).

> **The proof**, first for QED, and then for QCD and other non-Abelian gauge theories, that the needed cancellations *do* indeed occur to all orders is, of course, a highly technical matter. It represents a triumph of theoretical physics, to which many distinguished physicists made important contributions. Here, we simply rely on their result that, in the theories of interest, the needed cancellations *do* occur.

## 3.5. The Renormalized Result and Its Ambiguities

The coefficient $r_1$ in the last equation depends not only on the physical variables $Q,\theta$, as we would expect, but also on $\mu$. However, the physical quantity $\mathcal{R}$ must ultimately depend only on the physical variables $Q,\theta$: it cannot depend on $\mu$ and the other arbitrary choices, as is evident from the original "bare" form Eq. (3.9). In the renormalized result, what happens is that the $\mu$ dependence in the coefficients $r_i$ cancels with the $\mu$ dependence from the couplant $a$.

That fact is expressed by the RG equation

$$\left( \mu \frac{\partial}{\partial \mu} \bigg|_a + \beta(a) \frac{\partial}{\partial a} \right) \mathcal{R} = 0, \tag{3.13}$$

where $\beta(a) \equiv \mu da/d\mu$. (Note the slight change in the normalization of $\beta$ relative Eq. (2.39) of the previous chapter.) Here the first term picks up the $\mu$ dependence via the $r_i$ coefficients while the second term picks up the $\mu$ dependence via $a$.

From the discussion in the previous chapter, we could have anticipated that the answer would appear in a form that seemingly involves an arbitrary scale $\mu$. If we could solve the theory exactly, there would be no problem, since the exact $\mathcal{R}$ is exactly independent of $\mu$. The difficulty arises because, in practice, we can calculate only a finite number of terms in the perturbation series. Since the $\mu$-dependence cancellations occur *across* different orders, they are spoiled when we truncate the series, leaving the approximant dependent on $\mu$.

The same point applies to the "other choices" involved in the definition of $g_R$. The couplant $a$ and the perturbative coefficients $r_i$ will each depend on these choices, but the exact result cannot. If we had made different choices when defining $g_R$, our renormalized couplant, $a'$, would have been different; instead of Eq. (3.6) we would have had

$$a' = a_B(1 + a_B I' + \cdots), \tag{3.14}$$

where $I'$ has the same divergent part as $I$, but a different finite part. Thus, eliminating $a_B$ to express $a'$ in terms of $a$, we would find

$$a' = a(1 + v_1 a + \cdots), \tag{3.15}$$

where $v_1 = I' - I$ is finite, but otherwise arbitrary.

If we had used the primed RS in the calculation of $\mathcal{R}$ we would have obtained a result

$$\mathcal{R} = a'^{\mathrm{P}}(1 + r_1' a' + \cdots), \tag{3.16}$$

where $r_1'$ is a *different* finite coefficient. It is given by $r_1' = r_1 - \mathrm{P}v_1$, as is easily seen:

$$
\begin{aligned}
\mathcal{R} &= {a'}^{\mathrm{P}}(1 + r_1'a' + \cdots) \\
&= (a(1 + v_1 a + \cdots))^{\mathrm{P}}\,(1 + r_1'a + \cdots) \\
&= a^{\mathrm{P}}(1 + (r_1' + \mathrm{P}v_1)a + \cdots),
\end{aligned}
\tag{3.17}
$$

so that $r_1$ must be $r_1' + \mathrm{P}v_1$.

A basic issue, that can cause much confusion, is the following. A change in $\mu$ affects the second-order coefficient $r_1$, but a change of the "other choices" — the RP — can do so, too. However, there is only *one* degree-of-freedom at this order, corresponding to the single arbitrary coefficient $v_1$ in Eq. (3.15). The resolution of this issue can only be properly explained later on.

**In Chapter 6,** when we have properly defined the $\tilde{\Lambda}$ parameter that specifies the boundary condition to the $\beta$ function, we will find that $\tilde{\Lambda}$ is RP dependent — in just the right way to make $r_1$ dependent on RS *only* through the ratio $\mu/\tilde{\Lambda}$. Any "other choices" that leave this ratio unchanged will only affect the higher coefficients $r_2, \ldots$. Chapter 7 will explain how to parametrize those "other choices."

## 3.6. The Cancellation of Divergences, Revisited

Let us revisit Eq. (3.6), the expression for the renormalized couplant in terms of the bare one:

$$
a = a_B(1 + a_B I(\mu) + \cdots).
\tag{3.18}
$$

and ask how it can be that the divergent integral $I(\mu)$ depends on a mass-scale $\mu$. The issue is starkest in a massless theory, such as massless QCD, where there is no dimensionful parameter in the theory's Lagrangian: What is there to form a ratio of $\mu$ with? The answer is that $I(\mu)$ is a divergent integral and can only be given a meaning by "regularizing" the theory, and regularization always introduces a scale — the cutoff $M_{\mathrm{uv}}$. In an actual Feynman-diagram calculation, one would find that $I(\mu)$ has the following form, for large $M_{\text{uv}}$:

$$I(\mu) = b \ln(M_{\text{uv}}/\mu) + w + \text{negl.} \tag{3.19}$$

Here $b$ and $w$ are finite coefficients that are independent of $\mu$, and "negl." are negligible terms that vanish as $M_{\text{uv}} \to \infty$. Note that there is a logarithmic divergence as $M_{\text{uv}} \to \infty$, and the argument of the logarithm is necessarily a dimensionless ratio of $M_{\text{uv}}$ with $\mu$.

The fact that the $\mu$ dependence of $I$ is logarithmic could have been anticipated from Eq. (3.11), where $r_1 = J(Q, \theta) - PI(\mu)$ must end up as a function of the *ratio* of $Q/\mu$ once $M_{\text{uv}}$ has been taken to infinity. The key cancellation is $\ln(M_{\text{uv}}/\mu) - \ln(M_{\text{uv}}/Q) = \ln(Q/\mu)$. Hence, we can predict the form of $J(Q, \theta)$ to be

$$J(Q, \theta) = Pb \ln(M_{\text{uv}}/Q) + j(\theta) + \text{negl}'. \tag{3.20}$$

The coefficient $b$ above is, in fact, the leading coefficient of the $\beta$ function, as we can quickly see. Differentiating Eq. (3.19) gives

$$\mu \frac{dI}{d\mu} = -b, \tag{3.21}$$

so that differentiating Eq. (3.18) gives

$$\beta(a) \equiv \mu \frac{da}{d\mu} = a_B^2 \, \mu \frac{dI}{d\mu} + O(a_B^3)$$

$$= -ba_B^2(1 + O(a_B)) \tag{3.22}$$

$$= -ba^2(1 + O(a)), \tag{3.23}$$

where the last step uses $a_B = a(1 + O(a))$ to eliminate $a_B$ in favour of $a$, so that no reference to bare quantities remains. In higher orders we would find a perturbation series for $\beta$:

$$\beta(a) = -ba^2(1 + ca + c_2 a^2 + c_3 a^3 + \cdots), \tag{3.24}$$

with finite coefficients. The $\beta$ function represents the "anomalous" $\mu$ dependence of the renormalized couplant arising from the presence of ultraviolet divergences in the unrenormalized theory.

## 3.7. Dimensional Regularization

We have so far framed our discussion in terms of cutoff-type proce-
dures. Those regularization methods have the conceptual advantage,
for our purposes, of cleanly separating the regularization procedure,
and its technicalities, from the RS — the definition of $g_R$ and
hence of $a$ — which can be specified by "renormalization conditions"
independent of the particular regularization. However, there is no
doubt that for actual calculations, dimensional regularization, and
its variants, are far more convenient.

Dimensional regularization considers a generalization of the
theory to a non-integer space–time dimension $d = 4 - \epsilon$. (This
generalization is by no means unique, and a whole set of rather arbitrary conventions must be specified.) Removing the regularization
corresponds to taking the limit $\epsilon \to 0$. The divergences — that in
a cutoff regularization appeared as logarithms of the cutoff $M_{\text{uv}}$ —
now appear as poles in $\frac{1}{\epsilon}$. The unrenormalized perturbation series
for a physical quantity has the form

$$\mathcal{R} = a_B \left( 1 + \left( \frac{R_{11}}{\epsilon} + R_{10} \right) a_B + \left( \frac{R_{22}}{\epsilon^2} + \frac{R_{21}}{\epsilon} + R_{20} \right) a_B^2 + \cdots \right),$$

$$(3.25)$$

where the $R_{ij}$'s are various coefficients. ($R_{11}$ is $b$, in fact.)

At first sight the fact that the regularization parameter is now
the dimensionless $\epsilon$ would seem to undermine our argument, at the
start of the preceding section, that regularization always introduces
a mass-scale. However, a mass-scale really is involved because in $d$
dimensions the bare coupling constant $g_B$ in the Lagrangian has
dimensions $[\text{mass}]^{-d/2+2} = [\text{mass}]^{\epsilon/2}$. Hence, if the renormalized
couplant, $a$, is to be dimensionless we will need to write

$$a_B = \mu_0^\epsilon a(1 - Ia + \cdots), \qquad (3.26)$$

where $\mu_0$ is some "unit of mass." In the same way, in any dimensional-
regularization calculation one repeatedly makes the small-$\epsilon$ expansion of $k^\epsilon$, where $k$ is some momentum or other massive variable, as

$$\text{`` } k^\epsilon = 1 + \epsilon \ln k + O(\epsilon^2). \text{''} \qquad (3.27)$$

While this is perhaps mathematically valid, it makes no dimensional sense, and is really

$$k^\epsilon = \mu_0^\epsilon (1 + \epsilon \ln(k/\mu_0) + O(\epsilon^2)), \tag{3.28}$$

where, again, $\mu_0$ is some "unit of mass." If one defines $a(\mu)$ by some regularization-independent renormalization conditions, the $I(\mu)$ integral becomes a function of the ratio $\mu/\mu_0$, while $J(Q,\theta)$ becomes a function of $Q/\mu_0$, with $\mu_0$ cancelling out in $r_1$.

However, in the well-known renormalization scheme called "minimal subtraction" (MS) the story is a bit different. The MS scheme corresponds to calculating physical quantities, such as $\sigma$, in dimensional regularization (with some very specific choices of the details) and then simply discarding the pole terms $\frac{1}{\epsilon}, \frac{1}{\epsilon^2}, \ldots$. While this might seem to be a blind disregard of infinitely large quantities, it is in fact equivalent to an $a_B$-to-$a$ substitution with a definition of $a$ that involves *only* the pole terms, with no finite parts. An even more popular scheme, "modified minimal subtraction" or $\overline{\text{MS}}$, subtracts the $(\ln 4\pi - \gamma_E)$ terms, where $\gamma_E$ is Euler's constant, that naturally accompany each $\frac{1}{\epsilon}$ pole. (This procedure could be recast as a minimal subtraction in a modified dimensional regularization with a different convention about the generalization to $d$ dimensions.)

In such schemes the definition of the RS is entwined with the specification of the regularization procedure. There is nothing wrong with that, provided the conceptual distinction between regularization (which is removed in the end) and renormalization (the definition of the couplant) is kept in mind. One may dispense with the distinction between $\mu_0$ and $\mu$.

In the minimal subtraction framework, as 't Hooft has shown, there is a strong formal argument that the bare coupling constant is infinitesimal, of order $\epsilon$. (See Exercise 3.2.) The result can be expressed as

$$a_B = \mu_0^\epsilon \frac{\epsilon}{b}\left(1 + c\frac{\epsilon}{b}\ln\frac{\epsilon}{b} + O(\epsilon)\right). \tag{3.29}$$

Note that the scale parameter $\mu_0$ is really unspecified because changing it by a finite factor only affects the $O(\epsilon)$ correction term.

We can then understand how the apparently nonsensical "bare" series, Eq. (3.25), with its divergent series coefficients can actually be hiding a finite result. Indeed, the bare expansion is akin to the extreme limit of a very bad RS where the couplant is far too small, so that the series coefficients are far too big.

## 3.8. Perturbation Theory, the Cutoff → ∞ Limit, and Asymptotic Freedom

Perturbation theory in a QFT involves two limits. One is the limit in which the regularization is removed; that is, $\epsilon \to 0$ in dimensional regularization, or $M_{uv} \to \infty$ in cutoff-type regularizations. The other is the perturbative limit in which, formally, $a \to 0$. We should perhaps ask if these two limits commute or not.

The answer, we would argue, depends on whether the theory is asymptotically free or not; that is whether the coefficient $b$ in $\beta(a) = -ba^2 + \cdots$ is positive or not.

As we saw in Sec. 2.8 the leading-order form of the effective couplant is

$$a \sim \frac{1}{b\ln(Q/\Lambda)}. \tag{3.30}$$

Thus, for an asymptotically-free theory the good region for perturbation theory is at large energies, $Q \gg \Lambda$. The formal perturbative limit, $a \to 0$, corresponds to pretending that $\Lambda \to 0$, so that perturbation theory is then good at any $Q$. The two limits can be expected to commute since the scales $\Lambda$ and $M_{uv}$ and are being sent off in different directions; to 0 and ∞, respectively. See Fig. 3.3(a).

In non-asymptotically-free theories, however, the good region for perturbation theory is at low energies, $Q \ll \Lambda$. (Recall that $a$ is proportional to $g^2$ and should be positive, as well as small.) Thus, here the perturbative limit corresponds to sending both $M_{uv}$ and $\Lambda$ to ∞. See Fig. 3.3(b). For the "real" theory we should send $M_{uv} \to \infty$ first, but perturbation theory takes $\Lambda \to \infty$ first. Consequently, perturbation theory in non-asymptotically-free theories is problematic. Indeed for $\lambda\phi^4$ theory and QED (viewed as

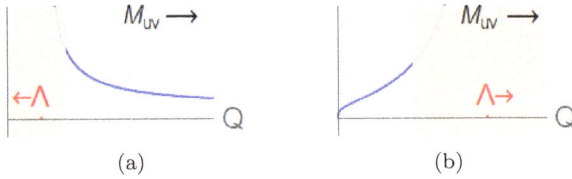

Fig. 3.3.   The leading-order effective couplant in (a) an asymptotically-free theory, and (b) a non-asymptotically-free theory. Perturbation theory is good in the unshaded region and becomes good at all $Q$ in the limit that the $\Lambda$ scale is taken to (a) zero or (b) infinity. In case (b) this limit conflicts with the cutoff to infinity limit, making perturbation theory dubious in non-asymptotically-free theories.

a self-contained theory) the "real" theory is believed to be "trivial" in the technical sense that it can have no interactions.

The present general attitude to perturbation theory in non-asymptotically-free theories is that it is valid provided that such theories are viewed as "effective low-energy theories" that approximate some unknown, grander theory with quite different ultraviolet behaviour that sets in at some finite, but very large, physical energy scale $M_{\mathrm{uv}}$. The regularization is then not some mere technical device, but is a crude representation of some actual physics at very high energies. (In terms of our "scaffolding" analogy, the scaffolding here is load-bearing, and it can't be entirely removed without causing the collapse of the whole structure.) Perturbation theory in this effective theory is meaningful provided that the $\Lambda$ scale is much greater than the $M_{\mathrm{uv}}$ scale. Of course, the uncertainties in what is the right regularization procedure, which now physically matters, are present alongside RS ambiguities in such theories. This book is principally concerned with asymptotically-free theories, where those issues do not arise.

## 3.9.   "RG-improved" Perturbation Theory and the RS-dependence Problem

Returning to the renormalized result for the physical quantity $\mathcal{R}$:

$$\mathcal{R} = a^{\mathrm{P}}(1 + r_1 a + \cdots),  \tag{3.31}$$

we now have a series with finite coefficients in a finite expansion parameter, $a$. However, the triumph of renormalization in rendering the result manifestly finite is tarnished by the fact that all the series coefficients $r_i$ are ambiguous and completely dependent on the arbitrary RS choices. That would not matter if all orders could somehow be calculated and re-summed, because $\mathcal{R}$ is RS invariant. However, there is a serious *RS-dependence problem* when we truncate the series, as we must in practice.

As noted earlier, the coefficient $r_1$ depends on $\mu/Q$ through a term $b \ln(\mu/Q)$. Thus, $r_1$ will have a very large magnitude if the physical scale $Q$ is much, much larger (or much, much smaller) than the renormalization scale $\mu$. The apparent convergence of the perturbation series would then be very bad. Because of this "large logarithm" problem, it is a bad idea to fix $\mu$ once-and-for-all; rather, it should be allowed to "run" with $Q$ in some fashion. Simply setting $\mu = Q$ leads to "RG-improved" perturbation theory, an idea originating with Gell Mann and Low and developed and exploited by the pioneers of QCD.

It is undoubtedly a major improvement — and was hugely important to the development of physics — but it does not resolve the ambiguities. Indeed, the idea of avoiding "large logarithms" only goes so far. Why should we set $\mu = Q$, rather than, say $\mu = 2Q$, or $\mu = Q/4$, etc.? Indeed, what, precisely, is "$Q$"? We have only said that $Q$ is some kinematic variable associated with the physical quantity $\mathcal{R}$, but which is the "right" one? Equally, we have said only that the renormalization scale $\mu$ is some parameter, with the dimensions of mass, introduced in the renormalization procedure. If one tries to address, separately, these three issues — what is the "right" $Q$, the "right" $\mu$, and the "right" relationship between them — one finds oneself sinking into a quagmire of ambiguities. This is not a useful way to approach the RS-dependence problem.

Moreover, the question of the "other choices" — the RP — would still remain. A frequent suggestion is to use a physical quantity to define a "physical" RP: Take some process $A$ with a normalized physical quantity $\mathcal{R}_A = a^{\mathrm{P}}(1 + \cdots)$ and define $a_A \equiv \mathcal{R}_A^{1/\mathrm{P}}$. One can then make predictions for another process $B$ as a truncated perturbation series in the "physical couplant," $a_A$. This is a superficially attractive idea, but it does not survive scrutiny. Why should we choose process $A$ as the special one? Why not processes $C, D, E, \ldots$? One is still left with an infinite number of differing predictions for process $B$. This is just the RS-dependence problem all over again: nothing has been resolved. (To say nothing of the issue raised above: How is the renormalization scale for process $B$ to be related to the "$Q$" of process $A$?) The definition of $a_A$ is just a way of specifying a RP. Indeed, the phrase "physical couplant" or "scheme-invariant couplant" is a contradiction in terms.

> **An analogy with Lorentz invariance** may help here. Energy, by its nature, is not an invariant quantity; it transforms from one reference frame to another. The rest mass of a particle *is* Lorentz invariant, and can be calculated from the energy $E$ and 3-momentum $\boldsymbol{p}$ of the particle in any frame, as the combination $\sqrt{E^2 - \boldsymbol{p}^2}$. Now, there is a particular frame, the rest frame, in which $\boldsymbol{p} = 0$ and $E = m$. Have we then discovered a "Lorentz-invariant energy"? Of course not: we have just defined a particular frame. In the same way, a couplant, by its nature, transforms under changes of RP. The definition of $a_A$ above does not define a "scheme-invariant couplant" — it defines a RP.

It is often tacitly assumed that the RP should be fixed once-and-for-all. In our view, that is a bad mistake. Just as $\mu$ should "run" with the energy scale $Q$, so the RP should be free to depend on the particular quantity $\mathcal{R}$ being calculated. This point should become clearer in the next chapter.

The reader may well find this section rather confusing and unsatisfactory. Indeed, its purpose is partly to give a flavour of the fog of confusion that too often surrounds the subject. The RS-dependence problem can only be discussed more productively when some fundamental matters — discussed in Chapters 6 and 7 — have been explained.

All these issues come into much better focus when RS-dependent perturbation theory is seen as a "non-invariant approximation." The quantum field theory itself does not have a problem — its exact results are RS independent. Thus, the answer is not to be found in the theory. The problem is with the approximation: The exact result is independent of the various RS variables, but the approximant depends upon those variables. Such "non-invariant approximations" actually occur in a wide variety of contexts, and much can be learned from some simple examples. That is the topic of the next chapter.

## Appendix 3.A: Renormalization in "Counterterm" Language

Renormalization is more usually described in "counterterm" language. For a renormalizable theory this just means that we make the reparametrization step in the Lagrangian from the start. To illustrate in the simplest case we consider a massless scalar theory with a $\phi^4$ interaction:

$$\mathcal{L} = \frac{1}{2}\partial_\mu \phi_B \, \partial^\mu \phi_B - \lambda_B \phi_B^4. \tag{3A.1}$$

This bare Lagrangian is written in terms of the bare field and bare coupling constant. ($\lambda_B$ is the counterpart to $a_B$ in the QCD case.) We now substitute for these in terms of a renormalized field and coupling constant:

$$\phi_B = Z_\phi^{1/2}\phi, \qquad \lambda_B = Z_\phi^{-2}Z_1\lambda. \tag{3A.2}$$

Here the $Z$'s are called renormalization constants and they will have a perturbation expansion of the generic from

$$Z = 1 + z_1\lambda + z_2\lambda^2 + \cdots, \tag{3A.3}$$

where the $z_1, z_2, \ldots$ coefficients are divergent integrals. With this substitution the Lagrangian becomes

$$\mathcal{L} = \frac{1}{2}Z_\phi \, \partial_\mu\phi \, \partial^\mu\phi - Z_1\lambda\phi^4, \tag{3A.4}$$

which we may then write as

$$\mathcal{L} = \frac{1}{2}\partial_\mu\phi \, \partial^\mu\phi - \lambda\phi^4 + \text{counterterms}. \tag{3A.5}$$

The counterterms have the same form as the terms in the "renormalized Lagrangian" but multiplied by $(Z_\phi - 1)$ and $(Z_1 - 1)$

factors, respectively, which start at order $\lambda$. From here on we need never mention the bare quantities. We may proceed to do Feynman-diagram perturbation theory treating both the $\lambda\phi^4$ term and the counterterms as perturbations. The counterterm coefficients $z_1, z_2, \ldots$ are to be fixed so that they cancel the divergences. (Of course, that only fixes the divergent parts of the $z_i$'s; the freedom to chose the finite parts corresponds to the RS ambiguity.)

This formalism has advantages in organizing the calculation. Green's functions with $n$ legs renormalize by

$$G^{(n)}(p_i, \lambda, \mu) = Z_\phi^{-n/2} G_B^{(n)}(p_i, \lambda_B, M_{uv}), \qquad (3A.6)$$

while proper vertices (1-particle irreducible amplitudes) renormalize with a $Z_\phi^{n/2}$ factor. For physical quantities the field normalization does not matter and only the composite $Z \equiv Z_\phi^{-2} Z_1$ in $\lambda_B = Z\lambda$ matters. (The counterterm formalism also has the advantage of generalizing to non-renormalizable theories: There the counterterms are not of the same form as the terms in the Lagrangian, and more and more are needed in higher orders.)

**Exercise 3.1.** Generalize the equations of Secs. 3.4–3.6 keeping *two* orders of correction terms and show that one can expect both $\ln^2(M_{uv}/\mu)$ and $\ln(M_{uv}/\mu)$ divergences, which — assuming the needed cancellations do happen — will leave $\ln^2(Q/\mu)$ and $\ln(Q/\mu)$ terms in the second-order coefficient $r_2$.

Show also that the second coefficient of the $\beta$ function can be found from the coefficient of the subleading $\ln(M_{uv}/\mu)$ divergence of the 2-loop term.

**Exercise 3.2.** This exercise follows the analysis of 't Hooft. In the minimal-subtraction scheme, if we group terms by powers of $1/\epsilon$, rather than by powers of $a$, the expression for the bare couplant in terms of the renormalized couplant is

$$a_B = \mu^\epsilon \left( a + \sum_{i=1}^\infty \frac{A_i(a)}{\epsilon^i} \right).$$

Here $a_B$ is $\mu$-independent, while $\mu\frac{da}{d\mu}$ will be $\beta(a)$ *plus corrections* of order $\epsilon$. Multiply through by $\mu^{-\epsilon}$ and take $\mu\, d/d\mu$ of both sides.

Then equate powers of $\epsilon$ to show that

$$\mu \frac{da}{d\mu} = \beta(a) - \epsilon a,$$

and that the $A_i(a)$'s are related to $\beta(a)$ by

$$\beta(a) = aA_1' - A_1,$$
$$\beta(a)A_j' = aA_{j+1}' - A_{j+1}, \quad j = 1, 2, \ldots.$$

Show that these results are equivalent to the equation

$$\frac{da_B}{da}(\beta(a) - \epsilon a) = -\epsilon a_B.$$

Integrate this differential equation to show that

$$a_B = \mu^\epsilon \exp\left(-\epsilon \int_\delta^a dx \frac{1}{\beta(x) - \epsilon x} - \int_\delta^1 \frac{dx}{x}\right),$$

with $\delta \to 0$. Alternatively, writing $\beta(x)$ as $-bx^2 B(x)$, one may express the result as

$$a_B = \mu^\epsilon a \exp\left(-b \int_0^a dx \frac{B(x)}{\epsilon + bxB(x)}\right).$$

Hence, noting that $B(x) \approx 1/(1 - cx)$ will be a sufficient approximation, obtain

$$a_B = \mu^\epsilon \frac{\epsilon}{b}\left(1 + c\frac{\epsilon}{b}\ln\frac{\epsilon}{b} + O(\epsilon)\right).$$

(Note that $\epsilon$ and $b$ need to have the same sign for $a_B$ to be positive.)

# Chapter 4

# Non-invariant Approximations and the Principle of Minimal Sensitivity

## 4.1. Approximations and Series Expansions

Before turning to non-invariant approximations specifically, some general consideration of approximations may be warranted. The discussion may seem insultingly elementary and our remarks rather trite, but we hope at least to convey that there are issues to be thought about. Theorists learn the mathematics of limits, asymptotics, conditions for convergence of series, etc., but these theorems, while immensely important, are not quite the same as the question of what makes a good approximation.

For instance, proving that the asymptotic behaviour of some function $A(Q)$ as $Q \to \infty$ is $A(Q) \sim \ln Q$ does not really answer the question *What is a good approximation to $A(Q)$ when the energy $Q$ is some specific, large value?* To answer the second question one must decide, at least roughly, what reference scale $Q_0$ belongs in the argument $Q/Q_0$ of the logarithm, which fundamentally must be dimensionless. The choice of $Q_0$ is irrelevant to the mathematical limit, but it matters quite a lot for the practical question. Similarly, answering the mathematical question of whether a series converges does not answer the question of whether, say, the first three terms of the series are likely to give a good approximation. Indeed, the second question may have very little to do with the first.

Approximations arise in many ways, but the justification always traces back to the fact that some dimensionless parameter is small. There is a limit — though it may or may not be realizable physically — in which the parameter tends to zero and the approximation becomes exact. In many cases the approximation can be systematically improved and we find ourselves calculating a quantity $f(x)$ as a power series,

$$f(x) = f_0 + f_1 x + f_2 x^2 + f_3 x^3 + \cdots , \qquad (4.1)$$

with our successive approximations being successive truncations of this infinite series. (Not all problems lead to such a form — the series could involve logarithms of $x$, or still more exotic apparitions — but, for simplicity, we will not discuss such cases.) When $x$ is small we can hope that $f_0$ is a crude approximation to $f(x)$, and that $f_0 + f_1 x$ is a better one, with $f_0 + f_1 x + f_2 x^2$ being better still, and so on.

A key mathematical issue is whether the series is *convergent*; that is, whether the partial sums $\sum_{i=0}^{n} f_i x^i$ converge to a limit as $n \to \infty$. Convergent series converge in a circular region of the complex $x$ plane, for $|x| < x_c$, where $x_c$ is the *radius of convergence*. Even if the series is not convergent it may be *asymptotic*. A prototypical example is the alternating factorial series:

$$f(x) = x(1 - 1!x + 2!x^2 - 3!x^3 + 4!x^4 + \cdots ). \qquad (4.2)$$

It has radius of convergence zero, since — no matter how small $x$ is — the partial sum eventually becomes dominated by the last term added, with the result then alternating violently between large positive and negative values. However, for $x$ small, the partial sums initially appear to converge towards a constant value (around 0.9156 for $x = 0.1$, for example). Thus we can often get useful, even remarkably accurate, approximations from divergent series.

It is important to distinguish two questions: *Is the series convergent?* (or, more generally, is it summable by such-and-such a method?) and *Is the sum of the series the right answer?* The first question depends only on the series, but the second cannot be addressed without knowing where the series comes from.

**For example,** consider the following fable. Suppose that the exact answer to some physical problem is given by $f(x)$ at $x = \frac{1}{5}$, and it is known that $f(x) \sim x$ as $x \to 0$. A complicated perturbative method is devised to calculate $f(x)$ as a power series. After long calculations the first few terms are found to be

$$f = x(1 - x + x^2 + \cdots).$$

A clever theorist then proves that the series coefficients, to all orders, are $\pm 1$ in alternation. Resumming the series he then obtains

$$f = \frac{x}{1 + x},$$

which gives $f(\frac{1}{5}) = 0.167$. Sadly, though, this result disagrees strongly with the experimental result, 0.433, even though the series is convergent, and $x = \frac{1}{5}$ is well within the radius of convergence.

Another clever theorist then establishes that $f(x)$ satisfies the differential equation

$$x^3(1 + x)\frac{d^2 f}{dx^2} - x(1 - 3x^2)\frac{df}{dx} + (1 - x)f = 0,$$

and that for large $x$ there is another boundary condition requiring $f \to 1 + 4\pi^2$ as $x \to \infty$. Thus, the actual answer is

$$f(x) = \frac{x}{1 + x} + 4\pi^2 e^{-\frac{1}{x}},$$

and the mystery is solved.

In fact the equal signs in the earlier equations were quite inappropriate: The first was not $f$ but the series generated from it, and the second was not $f$ but the sum of that (convergent) series. The actual $f$ has a large non-perturbative term, invisible to perturbation theory (the power series expansion of $e^{-\frac{1}{x}}$ is $0 + 0 + 0 + \cdots$). The moral of our fable is that the possible existence of such terms is an issue, *irrespective* of whether the series is convergent or divergent. Note that there is simply no information in the coefficients of the series about the non-perturbative term. No clever resummation method could have led, except by complete accident, to the true answer.

For divergent series, the question of how to sum the series and the possibility of non-perturbative terms are somewhat intertwined, since there can be a natural sense in which non-perturbative terms arise from the needed resummation procedure, their form being in a sense implied by the rate of growth of the coefficients. However, there are still two distinct issues, since there can always be non-perturbative terms that are just "tacked on" and about which the series coefficients know nothing. Thus, a proper discussion must start, not from the series, but from the mathematical problem (differential equation, functional integral, or whatever) from which the series arose. (See also Exercise 4.1.) Such issues will be relevant in Chapter 5. For now we focus more on approximations obtained from just a few terms of a series.

## 4.2. Approximology

Approximations are a gamble. In physics, as in life, we have to act upon incomplete information. Except in rare instances where rigorous inequalities can be proved, we can say nothing with certainty. Nevertheless, our duty is to obtain the best approximation we can, given limited time and resources. We should be clear, though, we are making a gamble.

There are, however, good gambles and bad gambles. The guiding principle must be to *make good use of any information we have.* One should not bet on an outcome that is glaringly inconsistent with some known information. Even the simplest facts can be powerful information.

**For example,** consider the classical ballistics problem of the range of a projectile with air resistance proportional to velocity. The projectile, mass $m$, is launched with a muzzle velocity $v_0$ at an angle of elevation $\alpha$. Neglecting air resistance the range is $R_0 = 2UV/g$ where $U = v_0 \cos \alpha$ and $V = v_0 \sin \alpha$ are the horizontal and vertical components of the initial velocity. Allowing for a drag force $\boldsymbol{F} = -K\boldsymbol{v}$ as a first-order perturbation gives an approximant to the range as

$$R \approx R_0 \left(1 - \tfrac{2}{3}\kappa\right),$$

where the dimensionless small parameter is

$$\kappa = \frac{2KV}{mg}.$$

This provides a satisfactory approximation only when $\kappa$ is much, much less than 1. Indeed, we can see that the approximant absurdly predicts a *negative* range if $\kappa > \frac{3}{2}$. The fact that the exact result *must* be positive definite should be made use of. A simple way to do so is to modify the form of our approximant to

$$R \approx \frac{R_0}{\left(1 + \frac{2}{3}\kappa\right)},$$

which leads to a much more robust approximation. This a simple example of a Padé approximant, which in general is the ratio of two polynomials in the small parameter. Padé approximants are not a universal panacea, but in this case there is a clear reason for preferring the Padé form.

Approximations are never without uncertainties and ambiguities. There are always two questions to be decided: *Which quantity do we want to calculate?* and *What form of approximant will we use?* The two questions are interlinked. For example, in the ballistics example above the choice of the Padé approximant form is equivalent to choosing to calculate the *reciprocal* of the range, rather than the range itself, as a power series. Another example is the following. Suppose a quantity $f$ has a power series $f_0 + f_1 x + f_2 x^2 + \cdots$, but there are reasons for thinking that $f^2$ is the more natural quantity. A second-order approximation to $f^2$ would be equivalent to using an approximant to $f$ of the form $(f_0^2 + 2 f_0 f_1 x + (2 f_0 f_2 + f_1^2) x^2)^{1/2}$ rather than $f_0 + f_1 x + f_2 x^2$. Similarly, if a quantity $f$ is naturally the ratio $g/h$ of two other quantities, do we want to calculate $f$ or to calculate $g$ and $h$ separately? This matters because the approximation to $f$ is, in general, not exactly the same as the ratio of the approximations to $g$ and $h$.

It is hard to give any general guidance on these issues, beyond saying that one should calculate what seems most natural and is most directly measured in experiments. The general principle should always be to make use of all information available in any specific case.

Discipline is needed when making approximations. If a calculation involves more than one approximation we should carefully distinguish the primary approximation — which is the unavoidable step and the main source of error — from secondary approximations that might be made just for convenience. Such secondary approximations are fine provided that they cannot contribute significantly to the overall error.

**In QCD,** for example, the primary approximation is the truncation of the perturbation series. A common practice is to use a secondary approximation where the couplant is approximated by a truncated series in $1/\ln(Q/\Lambda)$. However that introduces uncontrolled errors and unnecessary ambiguities, as will be discussed in Chapter 6.

It is always interesting to investigate approximations beyond their comfort zone. If an approximant becomes manifestly unphysical beyond some point then perhaps we are not making best use of our known information, and some modification would cure, or at least mitigate, the problem — as we saw in the ballistics example above. Even if the region of applicability of the approximation is intrinsically and unavoidably limited, it is good to know how and why it breaks down, if only to better appreciate its likely uncertainties in the intermediate region.

**For instance,** QCD perturbation theory is good at high energies, but is it necessarily useless at low energies? The effective couplant is guaranteed to be small when $Q \gg \Lambda$, but, in some cases at least, it may remain small even when $Q/\Lambda$ is small. That opens up the possibility — modulo non-perturbative terms — that perturbation theory may have something useful to say about low energies. (Note that a secondary approximation of re-expanding in powers of $1/\ln(Q/\Lambda)$ would spoil any such possibility.) This issue is intertwined with RS dependence, of course, so it is too early to discuss it further.

## 4.3. The Principle of Minimal Sensitivity

"Non-invariant approximations" are the focus of the remainder of the chapter. These are approximations where the results depend on some "extraneous" parameter(s) that we know the exact result cannot depend on. Non-invariant approximations can arise in various ways. Suppose — in the days before computers — a physicist needed to accurately evaluate an integral on the range 0 to 1, with an integrand that was singular at the bottom limit, behaving as, say, $1/\sqrt{x}$, as

$x \to 0$, but otherwise smooth. He or she might naturally make an asymptotic expansion about $x = 0$ and integrate that analytically up to some small value $x_0$, and then add the result to a numerical integration from $x_0$ to 1. But, what value of $x_0$ should be used? The answer is, of course, "it should not matter." However, while it *should not* matter, the uncomfortable fact is that it *does*. Of course, if one needed only two-decimal-place accuracy and the results, for a few plausible values of $x_0$ only differed in the fifth decimal place, then one would probably be quite happy. But if it were very important to get the best possible accuracy then it would make sense to think carefully about what $x_0$ is optimal. Our argument is that, in the absence of other information, the best choice for $x_0$ is where the result is least sensitive to small variations in $x_0$. We call this the "Principle of Minimal Sensitivity."

> *If an approximant depends on "extraneous" parameters, then — in the absence of further information — their values should be chosen so as to minimize the sensitivity of the approximant to small variations in those parameters.*

We mean this just as an explicit statement of a piece of common sense. It is a notion that has no doubt been employed many times in various specific contexts without much fanfare. By calling it a "principle" we do not mean to suggest that it is on a par with, say, the Principle of Least Action, but merely to emphasize its great generality. We regard it almost as a moral principle: "things should be independent of what things should be independent of."

There is no theorem here. We are talking about what is the best gamble to make, given all that we know. The fact that we *know* that the exact result is independent of $x_0$ is a valuable piece of information, and ought to be made use of. One could hardly believe that the approximation was a good one if it varied a lot under a small change of $x_0$. Our point is that where the approximate result is least sensitive to small variations of $x_0$ is where it is most believable. (See Fig. 4.1.)

Of course, the argument is qualified by the phrase "in the absence of further information," since in rare cases we might know a specific

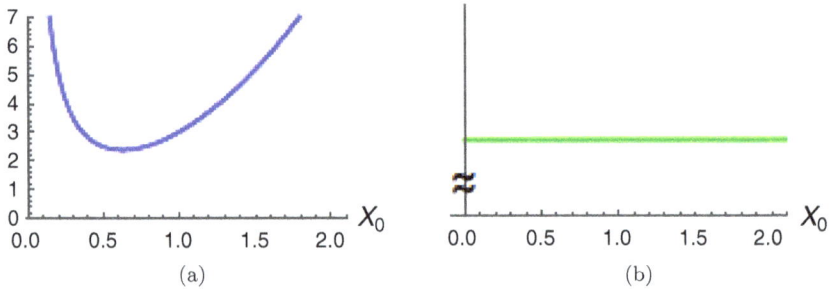

Fig. 4.1. (a) A non-invariant approximation gives a result that is a function of some extraneous parameter $x_0$. (b) We also know that the exact result, whatever its value may be, is independent of $x_0$. In what sense is (a) a good approximation to (b)?

fact that definitely indicated a different choice for the extraneous parameter. (Even in such cases, we would argue, it is likely that the alternative choice is quite close to the minimal-sensitivity choice, and makes only an insignificant difference.) In many cases it is possible to make intuitive arguments for a good choice of the extraneous parameter — an example is in Fig. 4.4 — but these arguments are generally very "fuzzy" and merely corroborate, and provide some insight into, the PMS choice.

> **It is possible,** of course, for the approximant to have more than one stationary point, or none. That issue does not arise in the RS-dependence problem, but it does in other applications of PMS. It will be discussed when it arises in some of the examples below.

To better understand the above arguments, and to see how non-invariant approximations work, one can learn a lot from some simple examples.

## 4.4. Quartic Oscillator Example: First-Order Approximation

Consider the classic quantum-mechanical problem of computing the eigenvalues, $E_n$, of an anharmonic oscillator. For maximal simplicity, we specialize to an oscillator with a purely quartic potential. (The general case, with both $x^2$ and $x^4$ terms in the potential will be discussed later in Sec. 4.8.) The Hamiltonian is

$$H = \tfrac{1}{2}p^2 + \lambda x^4, \tag{4.3}$$

where $x$ and $p$ are operators satisfying the commutation relation $[x, p] = i$. Trying to treat the $\lambda x^4$ term as a "perturbation" is quite hopeless — the unperturbed Hamiltonian would then be just $\tfrac{1}{2}p^2$, which has a continuous spectrum, qualitatively unlike the discrete spectrum of $H$. However, following the work of Caswell and Killingbeck, we may add and subtract an $x^2$ term and write $H$ as $H_0 + H_{\text{int}}$ with

$$H_0 = \tfrac{1}{2}\left(p^2 + \Omega^2 x^2\right), \quad H_{\text{int}} = -\tfrac{1}{2}\Omega^2 x^2 + \lambda x^4. \tag{4.4}$$

Standard quantum-mechanical perturbation theory can now be applied, generating what we call the Caswell–Killingbeck (CK) expansion (it is also known as the "linear $\delta$ expansion").

Note that an "extraneous" parameter, $\Omega$, has been introduced. It is arbitrary, in the sense that the exact eigenvalues of $H$ clearly do not depend on $\Omega$. However, our approximate results at any finite order will depend on $\Omega$; hence we describe the approximation as being "non-invariant."

**The calculation** is most easily done by introducing raising and lowering operators, $a^\dagger, a$, for a simple-harmonic oscillator of frequency $\Omega$. These have the commutation relation $[a, a^\dagger] = 1$ and are related to the $x, p$ operators by

$$x = \frac{1}{\sqrt{2\Omega}}(a + a^\dagger), \quad p = -i\sqrt{\frac{\Omega}{2}}(a - a^\dagger),$$

so that $H_0 = \Omega(a^\dagger a + \tfrac{1}{2})$. The unperturbed eigenvalues are

$$E_n^{(0)} = (n + \tfrac{1}{2})\Omega.$$

The unperturbed states are given by

$$|n\rangle^{(0)} = \frac{a^{\dagger n}}{\sqrt{n!}}|0\rangle^{(0)},$$

and satisfy

$$a^\dagger |n\rangle^{(0)} = \sqrt{n+1}|n+1\rangle^{(0)}.$$

Straightforward calculations using the commutation relations yield

$$^{(0)}\langle n|(a + a^\dagger)^2|n\rangle^{(0)} = 2n + 1,$$
$$^{(0)}\langle n|(a + a^\dagger)^4|n\rangle^{(0)} = 3(2n^2 + 2n + 1).$$

The first-order correction term for the $n$th eigenvalue is

$$E_n^{(1)} = {}^{(0)}\langle n|H_{\text{int}}|n\rangle^{(0)} = -\frac{1}{2}\left(n + \tfrac{1}{2}\right)\Omega + \frac{3\lambda}{4\Omega^2}(2n^2 + 2n + 1). \quad (4.5)$$

Adding this correction to the zeroth-order term gives the first-order result:

$$E_n^{\text{res}} = E_n^{(0)} + E_n^{(1)} = \frac{1}{2}(n + \tfrac{1}{2})\Omega + \frac{3\lambda}{4\Omega^2}(2n^2 + 2n + 1). \quad (4.6)$$

Note that the first term has a factor $\frac{1}{2}$ in front arising from a partial cancellation, $1 - \frac{1}{2}$, between the zeroth-order term and the part of the first-order correction produced by the $-\frac{1}{2}\Omega^2 x^2$ piece of $H_{\text{int}}$. Note also that $E_n^{\text{res}}$ can be written as $^{(0)}\langle n|H_0|n\rangle^{(0)} + {}^{(0)}\langle n|H_{\text{int}}|n\rangle^{(0)} = {}^{(0)}\langle n|H|n\rangle^{(0)}$, so that the first-order perturbative result can also be viewed as a variational estimate using the trial state $|n\rangle^{(0)}$.

The result (4.6) naturally depends on the physical variable $n$, the quantum number labelling the eigenstates, but it also depends on the extraneous variable $\Omega$. Because of this unphysical dependence the result has no quantitative meaning unless and until we decide how $\Omega$ is to be chosen. A plot of $E_n^{\text{res}}$ against $\Omega$, Fig. 4.2, shows that taking $\Omega$ too big or too small is clearly bad; the result becomes infinitely large in either limit. But how, in the absence of further information, are we to find the "right" value of $\Omega$? (And is it the same for all energy levels, or does it depend on $n$?)

We *do* have one piece of information, however: We know that the exact $n$th eigenvalue, $E_n$, is *independent* of $\Omega$. That is surely a valuable piece of information, and ought to be made use of. The Principle-of-Minimal-Sensitivity (PMS) argument is that it is sensible to choose $\Omega$ to be in a region where the approximant is insensitive to small variations of $\Omega$. (Both Caswell and Killingbeck independently made this argument.) The approximant, $E_n^{\text{res}}$, is

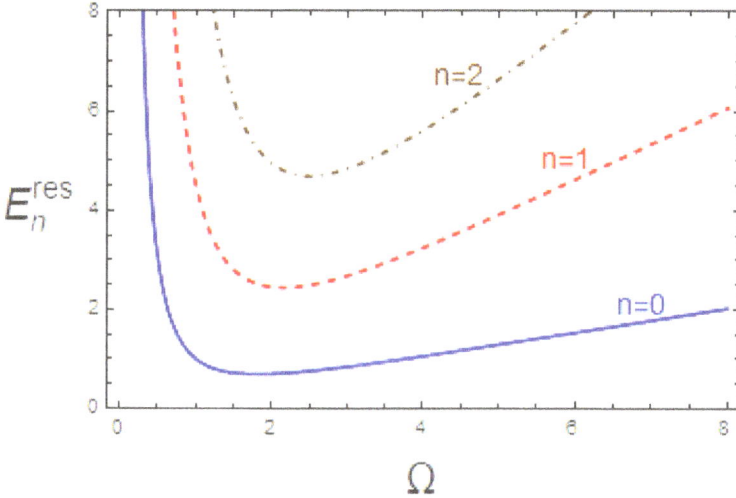

Fig. 4.2. Results for the low-lying energy levels $E_0, E_1, E_2$ to first order in the CK expansion, as a function of the extraneous variable $\Omega$, in units of $\lambda^{1/3}$.

minimally sensitive to $\Omega$ at the stationary point where

$$\frac{dE_n^{\mathrm{res}}}{d\Omega} = 0. \tag{4.7}$$

This is the "minimal sensitivity" or "optimization" condition. Its solution yields the "optimal" value of the $\Omega$ parameter — which indeed depends on which energy level we are calculating.

Applied to Eq. (4.6) the PMS optimization condition leads to the equation

$$\frac{1}{2}(n + \tfrac{1}{2}) + (-2)\frac{3\lambda}{4\Omega^3}(2n^2 + 2n + 1) = 0, \tag{4.8}$$

whose solution is

$$\bar{\Omega} = \left[ 3\lambda \frac{(2n^2 + 2n + 1)}{(n + \tfrac{1}{2})} \right]^{\frac{1}{3}}. \tag{4.9}$$

(Quite generally, we shall use an overbar to denote an "optimized" value.) Substituting this value into Eq. (4.6) yields the "optimized"

Table 4.1. The energy levels $E_n$ of the quartic oscillator in units of $\lambda^{1/3}$. The approximate eigenvalues $\bar{E}_n^{\rm res}$ from first order in the CK expansion, optimized by the PMS criterion, are compared with the essentially exact results from Hioe *et al.*

| $n$ | $\bar{E}_n^{\rm res}$ | $E_n^{\rm exact}$ | Error |
|---|---|---|---|
| 0 | 0.68142 | 0.667986 | 2.01% |
| 1 | 2.42374 | 2.39364 | 1.26% |
| 2 | 4.68500 | 4.69680 | −0.25% |
| 3 | 7.29111 | 7.33573 | −0.61% |
| 10 | 31.3587 | 31.6595 | −0.95% |
| $n \to \infty$ | $(1.36284)$ $\times (n + \frac{1}{2})^{4/3}$ | $(1.37651)$ $\times (n + \frac{1}{2})^{4/3}$ | −0.99% |

result:

$$\bar{E}_n^{\rm res} = \frac{3}{4}(n + \tfrac{1}{2})\left[3\lambda\frac{(2n^2 + 2n + 1)}{(n + \tfrac{1}{2})}\right]^{\frac{1}{3}}. \tag{4.10}$$

If we compare with precise values from the literature (see Table 4.1), we find that this simple formula fits *all* the energy levels to within 2%. The results for $n = 0$ and $n = 1$ are slight overestimates, by 2% and 1%, respectively, while for $n = 2$ and above the results are slight underestimates, by less than 1%. (For large $n$ one can compare with the semi-classical result from Bohr–Sommerfeld quantization.)

## 4.5. Quartic Oscillator Example: Discussion

The key to this success is the "optimal" choice of $\Omega$, which is different for different $n$. For any fixed ($n$-independent) value of $\Omega$, the approximate result, Eq. (4.6), would give a very poor description of the spectrum, so it is crucial that $\Omega$ is allowed to "run" with $n$. The PMS optimization naturally selects an appropriately $n$-dependent $\bar{\Omega}$.

Clearly, the PMS optimization condition (4.7) is reminiscent of the Variational Principle. However, only in a few specific instances there is a clear-cut connection. For the ground state ($n = 0$) our approximation corresponds to a variational calculation, with the ground state of the harmonic oscillator $H_0$, as the trial wavefunction.

In this instance we know, from the variational (Rayleigh–Ritz) theorem, that the approximate result $E_0^{\text{res}}$ is greater than or equal to the exact eigenvalue $E_0$ for any value of $\Omega$. Therefore by minimizing with respect to $\Omega$, as the PMS criterion does, we obtain an approximation to $E_0$ that is unquestionably optimal.

> **The same statement** can be made about the first-excited-state $(n = 1)$ case, using an extension of the variational theorem which says that the variational form $\langle \psi | H | \psi \rangle$ is greater than or equal to the exact $E_1$ for all states $|\psi\rangle$ that are orthogonal to the exact ground state. Our trial state $|1\rangle^{(0)}$ is guaranteed to be orthogonal to the unknown exact ground state because the former is odd under parity, $x \rightarrow -x$, whereas the latter is even.

For $n \geq 2$, however, there is no variational inequality guaranteeing that the approximate result must be greater than the exact eigenvalue for all values of $\Omega$. Indeed, such an inequality does not hold; it is violated in a small region of $\Omega$ around the $\bar{\Omega}$ value. (See Fig. 4.3).

Thus, for $n \geq 2$, the PMS choice of $\Omega$ is not optimal in the rigorous sense of minimizing the error; there are other values of $\Omega$ that would lead to even more accurate results. Indeed, there are two "magic" values of $\Omega$, either side of the PMS $\bar{\Omega}$, that would lead to the *exact* eigenvalue. Finding those values, however, would

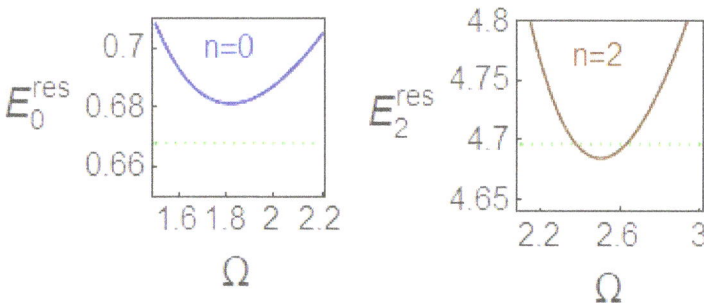

Fig. 4.3. Close-ups of the region around the optimal $\Omega$, showing the comparison with the exact result. For the ground state, $n = 0$ the variational inequality guarantees that the first-order approximant is always greater than the exact result. For $n = 2$ and higher there is no such inequality. Nevertheless, in both cases the PMS choice of $\Omega$ gives an approximation of comparable accuracy.

require some real magic, since that would be equivalent to solving the quartic-oscillator problem exactly. When we refer to the PMS choice as "optimal" we clearly cannot — except in special circumstances, such as the $n = 0, 1$ cases — claim to be using the word in a mathematically rigorous way. We mean it in the sense of "the best choice, in the absence of further information."

An intuitive understanding of why and how $\bar{\Omega}$ depends on $n$ can be gained from the following rough argument. For a good approximation, we want our approximate potential energy, $\frac{1}{2}\Omega^2 x^2$ to be a good approximation to the actual potential energy, $\lambda x^4$. Obviously, this cannot be true for all $x$, but what we most need, when considering the $n$th energy level, is for it to be true where both potential energies are of order of the unperturbed energy $E_n^{(0)}$. (See Fig. 4.4.) That is we want

$$(n + \tfrac{1}{2})\Omega \approx \tfrac{1}{2}\Omega^2 x^2 \approx \lambda x^4. \tag{4.11}$$

From the above approximate relations, we can eliminate $x$ as $x^2 \approx \frac{1}{2}\Omega^2/\lambda$ and solve for $\Omega$:

$$\Omega \approx (4(n + \tfrac{1}{2})\lambda)^{1/3}. \tag{4.12}$$

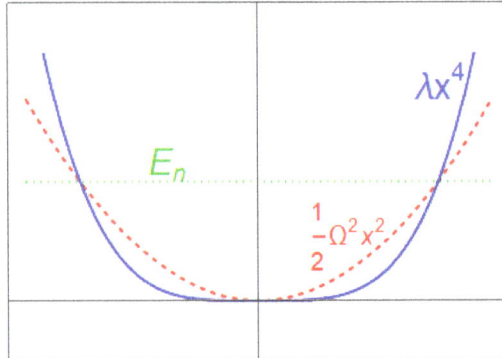

Fig. 4.4.    A sketch illustrating an intuitive argument for why $\Omega$ should increase with $n$: We wish the zeroth-order potential $\frac{1}{2}\Omega^2 x^2$ to be a good approximation to the actual potential, $\lambda x^4$ for energies of order $E_n$. For higher $E_n$ we will need larger $\Omega$, corresponding to a steeper quadratic potential.

This is in rough accord with Eq. (4.9), having the crucial $(n + \frac{1}{2})^{1/3}$ behaviour for large $n$.

Note that this is a hand-waving, order-of-magnitude argument; the details and the numerical factors are all debatable. (It is similar to the argument, in the RS-dependence problem, that the renormalization scale $\mu$ should be chosen to be of order the physical scale $Q$. In both cases the argument is correct; good physics; but inherently vague.) Trying to use such "physical" arguments alone to fix the value of extraneous parameters is never satisfactory in our experience. However, they often provide physical insight into why the PMS optimized value is what it is.

**Some authors,** faced with a non-invariant approximation, choose to fix the extraneous parameter, not by PMS, but by a notion that we call "fastest apparent convergence" (FAC). That approach is critically discussed in Appendix 4.A.

## 4.6. Quartic Oscillator Example: Second Order

One may proceed to calculate the energy eigenvalues to the next order in this perturbation theory. For simplicity, we consider only the ground state, $n = 0$. The general formula for the second-order correction to the ground-state energy is

$$E_0^{(2)} = \sum_{j \neq 0} \frac{|\,^{(0)}\langle j|H_{\text{int}}|0\rangle^{(0)}\,|^2}{E_0^{(0)} - E_j^{(0)}}, \tag{4.13}$$

where the sum runs over all the states except $j = 0$. In the quartic-oscillator case, the only non-zero contributions are from $j = 2$ and $j = 4$.

**From the** $a, a^\dagger$ algebra one finds that

$$^{(0)}\langle 2|x^2|0\rangle^{(0)} = \frac{\sqrt{2}}{2\Omega}, \quad ^{(0)}\langle 2|x^4|0\rangle^{(0)} = \frac{6\sqrt{2}}{(2\Omega)^2},$$

$$^{(0)}\langle 4|x^4|0\rangle^{(0)} = \frac{\sqrt{24}}{(2\Omega)^2},$$

and from these results one obtains the two terms, which are

$$\frac{|\,^{(0)}\langle 2|H_{\mathrm{int}}|0\rangle^{(0)}\,|^2}{E_0^{(0)} - E_2^{(0)}} = \frac{1}{(-2\Omega)}\left(-\tfrac{1}{2}\Omega^2\left(\frac{\sqrt{2}}{2\Omega}\right) + \lambda\left(\frac{6\sqrt{2}}{(2\Omega)^2}\right)\right)^2,$$

$$\frac{|\,^{(0)}\langle 4|H_{\mathrm{int}}|0\rangle^{(0)}\,|^2}{E_0^{(0)} - E_4^{(0)}} = \frac{1}{(-4\Omega)}\lambda^2\left(\frac{\sqrt{24}}{(2\Omega)^2}\right)^2.$$

Hence, the correction is

$$E_0^{(2)} = -\frac{\Omega}{16}\left(1 - \frac{6\lambda}{\Omega^3}\right)^2 - \frac{3}{8}\Omega\left(\frac{\lambda}{\Omega^3}\right)^2.$$

This must be added to the result from first order, from Eq. (4.6) with $n = 0$.

The second-order result is

$$E_0^{\mathrm{res}[2]} = \frac{3}{16}\Omega\left(1 + 8\left(\frac{\lambda}{\Omega^3}\right) - 14\left(\frac{\lambda}{\Omega^3}\right)^2\right). \qquad (4.14)$$

As a function of $\Omega$ this result does not have a stationary point, but it does exhibit a flat region around a point of inflexion. See the dashed curve in Fig. 4.5.

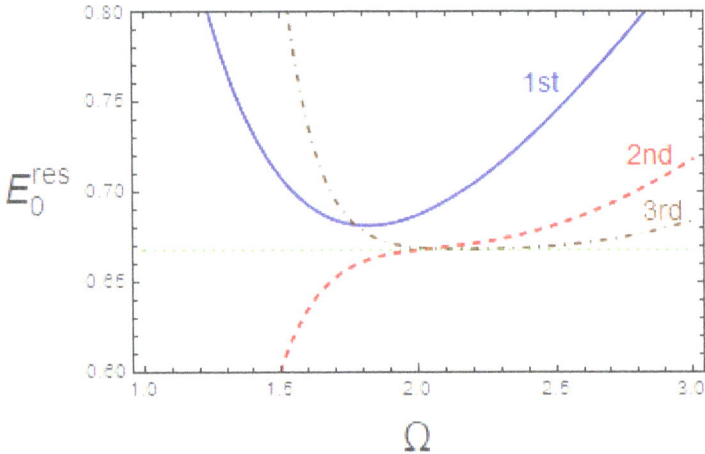

Fig. 4.5.   Results for $E_0$ to first, second, and third orders in the CK expansion, as a function of the extraneous variable $\Omega$.

In such cases, which do not arise in the RS-dependence problem, the application of PMS is not entirely unambiguous. There are two reasonable alternative strategies. (a) One may use the point of inflexion, where

$$\frac{d^2 E_0^{\text{res}}}{d\Omega^2} = 0. \tag{4.15}$$

The argument here is that, while we cannot make the slope, $dE_0^{\text{res}}/d\Omega$, zero, we can find where the slope is minimized, so that the result is "least sensitive to small variations." A somewhat unsatisfactory aspect of this argument is that the result then depends on selecting $\Omega$ as the extraneous parameter, rather than, say $1/\Omega$, or $\Omega^3$, or some other function of $\Omega$. The usual PMS criterion, Eq. (4.7), has the nice property of being invariant under such re-definitions, but the second-derivative does not share that property. However, this concern is minor, in that, for any remotely reasonable choices of variable, the result changes only by an amount well within the error that one would estimate anyway. (b) The other strategy is to allow complex solutions to the usual PMS condition. Close to the point of inflexion there will be a complex-conjugate pair of roots to Eq. (4.7). Evaluating the result at either of these roots will yield an answer with a small imaginary part, which we can then drop. We could view this as averaging the results at the two roots, since they will have equal and opposite imaginary parts. In doing so we are making use of another piece of information we have about the exact result; namely, that it is real.

The results of methods (a) and (b) are given in Table 4.2. They agree quite closely, and each produces a gratifying improvement on

Table 4.2.  Optimized results, in first and second order of the CK expansion, for the ground-state energy of the quartic oscillator, in units of $\lambda^{1/3}$. The exact value is 0.66798626 ....

| Order | $\bar{\Omega}$ | $\bar{E}_0^{\text{res}}$ | Error |
|---|---|---|---|
| 1st | 1.81712 | 0.681420 | 2.0% |
| 2nd (a) | 2.06064 | 0.668973 | 0.15% |
| 2nd (b) | $2.02015 \pm 0.20073i$ | $0.668641 \pm 0.002191i$ | 0.10% |

the first-order result, reducing the error from about 2% to less than 0.2%.

We will return in the next chapter to discuss higher orders in the CK expansion.

## 4.7. Quartic Oscillator Example: Wavefunctions

One may also obtain accurate approximate wavefunctions with the same approach, as shown by Kauffmann and Perez (KP), whose work we follow in this section. The key point is this: Since the wavefunction is not a single number, but a function of position, $x$, the "optimal" $\Omega$ will depend on $x$. Here, we will consider only the ground-state wavefunction, but the story is similar for the excited states.

**The calculation** proceeds as follows: The formula for the state vector of the $n$th eigenstate, to first order in perturbation theory, is

$$|n\rangle = |n\rangle^{(0)} + \sum_{j \neq n} c_{nj} |j\rangle^{(0)},$$

with

$$c_{nj} = \frac{{}^{(0)}\langle n|H_{\text{int}}|j\rangle^{(0)}}{E_n^{(0)} - E_j^{(0)}}.$$

In the case of the ground state, $n = 0$, the only non-zero terms in the sum come from $j = 2$ and $j = 4$. From the results in the previous section, one can obtain the two non-zero coefficients, which are

$$c_{02} \equiv \frac{{}^{(0)}\langle 2|H_{\text{int}}|0\rangle^{(0)}}{E_0^{(0)} - E_2^{(0)}} = \frac{1}{(-2\Omega)}\left(-\tfrac{1}{2}\Omega^2\left(\frac{\sqrt{2}}{2\Omega}\right) + \lambda\left(\frac{6\sqrt{2}}{(2\Omega)^2}\right)\right),$$

$$c_{04} \equiv \frac{{}^{(0)}\langle 4|H_{\text{int}}|0\rangle^{(0)}}{E_0^{(0)} - E_4^{(0)}} = \frac{1}{(-4\Omega)}\lambda\left(\frac{\sqrt{24}}{(2\Omega)^2}\right).$$

The Schrödinger wavefunctions representing the unperturbed states $|0\rangle^{(0)}, |2\rangle^{(0)}, |4\rangle^{(0)}$ are

$$\psi_0^{(0)} = \left(\frac{\Omega}{\pi}\right)^{1/4} \exp\left(-\tfrac{1}{2}\Omega x^2\right),$$

$$\psi_2^{(0)} = \left(\frac{\Omega}{\pi}\right)^{1/4} \exp\left(-\tfrac{1}{2}\Omega x^2\right) \frac{1}{\sqrt{2}} \left(-1 + 2\Omega x^2\right),$$

$$\psi_4^{(0)} = \left(\frac{\Omega}{\pi}\right)^{1/4} \exp\left(-\tfrac{1}{2}\Omega x^2\right) \frac{1}{2\sqrt{6}} \left(3 - 12\Omega x^2 + 4\Omega^2 x^4\right).$$

The result for the ground-state wavefunction, to first order in perturbation theory is then

$$\psi_0^{\text{res}} = \psi_0^{(0)} + c_{02}\psi_2^{(0)} + c_{04}\psi_4^{(0)}.$$

Note that the overall normalization will need to be adjusted at the end of the calculation.

The result of the perturbative calculation is

$$\psi_0^{\text{res}} = \left(\frac{\Omega}{\pi}\right)^{1/4} \exp\left(-\tfrac{1}{2}\Omega x^2\right) \frac{1}{16}$$

$$\times \left(14 + 4\Omega x^2 + \frac{\lambda}{\Omega^3}(9 - 12\Omega x^2 - 4\Omega^2 x^4)\right). \tag{4.16}$$

Optimizing $\Omega$, requiring $\partial \psi_0^{\text{res}} / \partial \Omega = 0$ at each $x$, leads to a quintic equation for the optimal $\Omega$:

$$2\Omega^3(7 - 4\Omega x^2 - 4\Omega^2 x^4) + \lambda(-99 + 66\Omega x^2$$

$$+ 36\Omega^2 x^4 + 8\Omega^3 x^6) = 0. \tag{4.17}$$

Picking the right root of this equation requires a little thought, as we now discuss (see Fig. 4.6).

At $x = 0$ the unique real root is $\bar{\Omega} = \left(\frac{99}{14}\lambda\right)^{1/3}$ — similar to, but not the same as, the optimum $\Omega$ for the ground-state energy, which was $(6\lambda)^{1/3}$. For small but non-zero $|x|$ there are two real, positive roots, but one goes to infinity as $x \to 0$, so clearly, for continuity, we want the smaller root that starts from $\left(\frac{99}{14}\lambda\right)^{1/3}$ at $x = 0$. This root persists until $x \approx 0.65\lambda^{-1/6}$, where it meets up with the other, larger root.

At large $|x|$ there are also two positive, real roots, but one goes to zero proportional to $1/x^2$: That would produce a non-normalizable form for $\psi$, and so can be rejected. The relevant root has $\bar{\Omega} \sim \sqrt{\lambda}\,|x|$ at large $|x|$ — which means that the resulting wavefunction has the

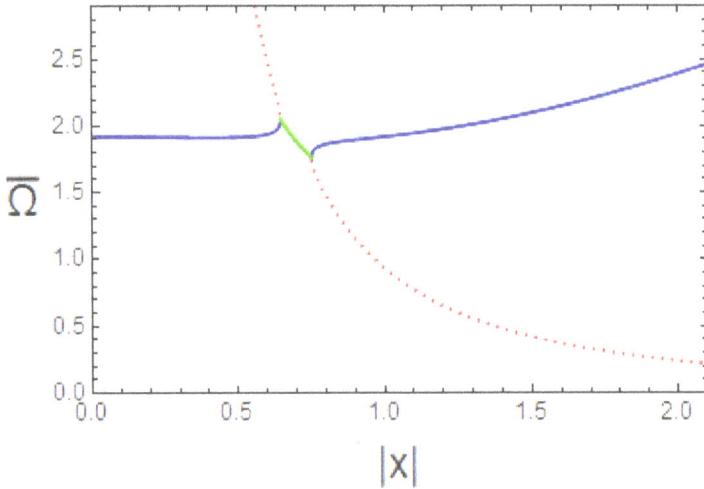

Fig. 4.6.    The optimal value of $\Omega$ as a function of $|x|$, in units where $\lambda = 1$. The relevant root is shown by the solid lines. The segment around $|x| \approx 0.7$ is actually the real part of a pair of complex roots. The three other roots, at any given $|x|$, are negative or have negative real parts. These, and the roots shown by the dotted lines, may be discarded as clearly inappropriate.

correct faster-than-Gaussian fall off, with an $\exp(-\frac{1}{2}\sqrt{\lambda}\,|x|^3)$ factor. By continuity, we can follow this root back to $x \approx 0.75\lambda^{-1/6}$, where it meets up with the smaller root.

There is region, roughly $0.65 < \lambda^{1/6}\,|x| < 0.75$, where there is no real, positive root. For $|x|$ in this range there is no actual stationary point in $\Omega$, only a flat region around where $\partial^2\psi/\partial\Omega^2$ vanishes. As discussed above, we can either use this point of inflexion, or — perhaps better — use either of the complex roots (whose real part is shown in the figure) and then, at the end, discard the tiny imaginary part of the resulting wavefunction.

Although the resulting $\bar{\Omega}$ as a function of $|x|$ has an odd-looking zig-zag (see Fig. 4.6), there is no discernible lack of smoothness in the resulting wavefunction. This is because $\psi$ is, of course, very insensitive to the precise value of $\Omega$, provided that it is close to $\bar{\Omega}$. The resulting optimized wavefunction, plotted either on a linear or a logarithmic scale, is hardly distinguishable from the exact wavefunction. (See figures in KP.) Nevertheless, it would be sensible to use an $\Omega$ that smooths out the zig-zag region of $\bar{\Omega}$, since we know that the exact wavefunction must be smooth.

## 4.8. Anharmonic Oscillator and Double-Well Potential

The more general case of an anharmonic oscillator — a simple harmonic oscillator of unperturbed frequency $\omega$, with an $x^4$ perturbation — can be treated very similarly. The Hamiltonian,

$$H = \tfrac{1}{2}p^2 + \tfrac{1}{2}\omega^2 x^2 + \lambda x^4, \tag{4.18}$$

can be written as $H_0 + H_{\text{int}}$ with

$$H_0 = \tfrac{1}{2}\left(p^2 + \Omega^2 x^2\right), \qquad H_{\text{int}} = -\tfrac{1}{2}\left(\Omega^2 - \omega^2\right)x^2 + \lambda x^4. \tag{4.19}$$

A simple generalization of the previous calculation yields the first-order correction term as

$$E_n^{(1)} = -\frac{1}{2}\left(n + \tfrac{1}{2}\right)\frac{\left(\Omega^2 - \omega^2\right)}{\Omega} + \frac{3\lambda}{4\Omega^2}(2n^2 + 2n + 1). \tag{4.20}$$

Added to $E_n^{(0)}$, this yields the first-order result

$$E_n^{\text{res}} = \frac{1}{2}\left(n + \tfrac{1}{2}\right)\frac{\left(\Omega^2 + \omega^2\right)}{\Omega} + \frac{3\lambda}{4\Omega^2}(2n^2 + 2n + 1). \tag{4.21}$$

Applying the PMS condition $dE_n^{\text{res}}/d\Omega = 0$ leads now to a cubic equation:

$$\left(n + \tfrac{1}{2}\right)\Omega\left(\Omega^2 - \omega^2\right) - 3\lambda(2n^2 + 2n + 1) = 0, \tag{4.22}$$

whose positive root is the optimal value, $\bar{\Omega}$. Substituting into $E_n^{\text{res}}$ yields the optimized result.

    **Caswell** has a nice formalism using a variable $\beta$ defined by

$$\Omega^2 = \omega^2 + \frac{\beta\lambda}{\Omega}.$$

which allows a unified treatment of all the cases $\omega^2$ positive, zero, or negative. ($\beta = 6$, which is the optimal value for the ground state at first order, corresponds to the mass renormalization induced by normal ordering the Hamiltonian.)

For all positive $w$ there is not much to say; the results are even better than for the quartic-oscillator case. Indeed, the larger $w$ is, relative to $\lambda^{1/3}$, the better the perturbation theory gets, and the less the optimal $\Omega$ differs from $w$. Nevertheless, the CK approach still has significant benefits over ordinary perturbation theory, particularly when one considers high-$n$ energy levels.

For imaginary $w$ (negative $w^2$) the classical potential has a double-well form. For $|w|$ small, relative to $\lambda^{1/3}$, the CK expansion remains good. However, for larger $|w|$ the CK expansion does not give satisfactory results — unless we are prepared to go to very high orders. There is a simple cure, however. We may generalize the method by adding and subtracting a linear term in $x$: That is, one takes $H_0$ to be an oscillator of frequency $\Omega$ whose equilibrium is shifted to some value $x_0$:

$$
\begin{aligned}
H_0 &= \tfrac{1}{2}\left(p^2 + \Omega^2(x - x_0)^2\right), \\
H_{\text{int}} &= -\tfrac{1}{2}\Omega^2(x - x_0)^2 + \tfrac{1}{2}w^2 x^2 + \lambda x^4.
\end{aligned}
\tag{4.23}
$$

There are now two extraneous parameters, $\Omega$ and $x_0$, that can be optimized by the PMS condition.

If we calculate the ground-state energy $E_0(x_0, \Omega)$ and then optimize $\Omega$, we obtain $\bar{V}_G(x_0)$, a function of $x_0$ that, for sufficiently negative $w^2$, has a local maximum at $x_0 = 0$ and a pair of minima at $x_0 = \pm\bar{x}_0$ for some non-zero $\bar{x}_0$. The optimized result for the ground-state energy $E_0^{\text{res}}$ is then given by $\bar{V}_G$ at $x_0 = \bar{x}_0$.

**The function** $\bar{V}_G(x_0)$ is called the "Gaussian effective potential." It can be calculated for a wide variety of problems $H = \tfrac{1}{2}p^2 + V(x)$ with all sorts of classical potentials $V(x)$ and it provides a good "picture" of how quantum zero-point-energy effects modify the physics — in the present case, for instance, it shows that, for $w^2$ negative but small, quantum effects wash out the hump in the classical potential. Moreover, the concept generalizes naturally to QFT, where it has significant advantages over the traditional one-loop effective potential. See citations in the references for this chapter.

One may similarly calculate results for the excited states. As $n$ increases the optimal $x_0$ slowly decreases, until at some sufficiently large $n$ there ceases to be a non-zero $\bar{x}_0$ and the optimized result comes from $x_0 = 0$.

## 4.9.  Conclusions

Approximations are a gamble; we are forced to act upon incomplete information. It is important to make good use of any and all information that we do have. Ambiguity is inherent in approximation; we always have to decide, somewhat arbitrarily, precisely what to calculate, and what the form of the approximant is to be.

Non-invariant approximations bring in another ambiguity; what value(s) should the extraneous parameter(s) take? The PMS criterion, once we have decided what we are calculating and what the form of the approximant is, provides an objective resolution of that ambiguity. It uses the information that the exact result is known to be exactly independent of the extraneous parameter(s).

The worst mistake with a non-invariant approximation is to assume that the extraneous parameter(s) have to be fixed once-and-for-all. In the quartic oscillator example, for example, one can get good results for all the energy levels — but only if the value of $\Omega$ is different for the different levels. The PMS criterion automatically selects a suitable $\Omega$ value for each case.

We are now ready to tackle the RS-dependence problem by viewing it as a non-invariant approximation and applying PMS optimization. The impatient reader may wish to skip the next chapter and proceed straight to Part II. The topic of the next chapter is high orders in examples of non-invariant approximations — the important point being that the optimal parameter(s) change from one order to the next, which can give rise to "induced convergence."

## Appendix 4.A: FAC Criteria

A different approach to dealing with non-invariant approximations is a notion dubbed "fastest apparent convergence" (FAC). The idea is that, at next-to-leading order, we would obviously *like* the correction term to be small in comparison to the leading term — so let us choose the extraneous parameter to make it so. Often, in fact, the correction term can be made to *vanish*. (Because the zeroth- and first-order results then agree exactly with each other, this choice is sometimes called a "self-consistency" criterion — but that terminology is misleading since there is no logical necessity.)

The FAC approach shares some of the virtues of the PMS method — crucially it allows the extraneous parameter to be different in different cases. Sometimes FAC and PMS give quite similar results, since the curves for two different orders will often intersect in the flat region. However, the curves can be nearly parallel in the flat region and bend away together, intersecting only quite far away, giving a poor result. In the author's opinion and experience, FAC is unreliable. It is predicated on a property that one would *like* the approximation to have — rapid apparent convergence — whereas PMS is based on a property that the exact result *does* have — invariance under variations of the extraneous parameter.

FAC is certainly inferior in the case of the CK expansion for the quartic oscillator. Requiring the correction term $E_n^{(1)}$, Eq. (4.5), to vanish gives the equation

$$-\frac{1}{2}\left(n+\tfrac{1}{2}\right)\Omega + \frac{3\lambda}{4\Omega^2}\left(2n^2 + 2n + 1\right) = 0, \qquad (4\text{A}.1)$$

whose solution is $\Omega_{\text{FAC}}$:

$$\Omega_{\text{FAC}} = \left[\frac{3\lambda}{2}\frac{\left(2n^2 + 2n + 1\right)}{\left(n+\tfrac{1}{2}\right)}\right]^{\frac{1}{3}}, \qquad (4\text{A}.2)$$

which is smaller than the PMS $\bar{\Omega}$ by a factor of $2^{-1/3}$. The FAC result is obtained by substituting this value of $\Omega$ into $E_n^{\text{res}}$, which, by construction, is the same as $E_n^{(0)} = (n+\tfrac{1}{2})\Omega$. Hence, the FAC result is

$$E_n^{\text{res}}|_{\text{FAC}} = \left(n+\tfrac{1}{2}\right)\left[\frac{3\lambda}{2}\frac{\left(2n^2 + 2n + 1\right)}{\left(n+\tfrac{1}{2}\right)}\right]^{\frac{1}{3}}. \qquad (4\text{A}.3)$$

It is very similar to the PMS result in form and for all $n$ values it is larger by a factor of $\frac{4}{3}2^{-1/3} \approx 1.06$. Consequently, it is a much less accurate approximation. See Table 4.3.

Table 4.3. FAC results, to first order in the CK expansion, for the energy levels $E_n$ of the quartic oscillator in units of $\lambda^{1/3}$. Compare with PMS results in Table 4.1.

| $n$ | $E_n^{\text{FAC}}$ | $E_n^{\text{exact}}$ | error |
|---|---|---|---|
| 0 | 0.721 | 0.667986 | 8.0% |
| 1 | 2.565 | 2.39364 | 7.2% |
| 2 | 4.958 | 4.69680 | 5.6% |
| 3 | 7.716 | 7.33573 | 5.2% |
| 10 | 33.186 | 31.6595 | 4.8% |
| $n \to \infty$ | $(1.442)$ | $(1.37651)$ | 4.8% |
| | $\times (n + \frac{1}{2})^{4/3}$ | $\times (n + \frac{1}{2})^{4/3}$ | |

At higher orders there are different ways to interpret the FAC idea: one could require either (i) that the last correction term vanishes so that adjacent orders agree, or (ii) that the net correction vanishes so that $k$th order agrees with zeroth order. In some examples the first approach is better than the second; in other examples it is the reverse. In the CK expansion method (ii) is hopelessly bad, while in Exercise 4.2 it is method (i) that is poor.

In the RS-dependence problem, because there are multiple extraneous parameters, it is possible to the choose the RS in $(k+1)$th order so that all the perturbative coefficients $r_1$ to $r_k$ vanish. Several authors have advocated use of this FAC or "effective charge" (EC) scheme. It is certainly useful for formal purposes (see Chapters 7 and 13), but it is doubtful that formal simplicity is a sound argument that it provides the best approximation. At low orders there is often very little difference between the FAC/EC and PMS results. However, the "induced convergence" scenario (to be discussed in the next chapter), in which $\mathcal{R}^{(k+1)}$ tends to a finite limit as $k \to \infty$ because $a \to 0$, clearly could not work for FAC/EC, where $a \to 0$ would entail $\mathcal{R} \to 0$.

**Exercise 4.1.** Consider the spatial integral

$$\int d^3r \left( \frac{2}{a_0^{3/2}} \frac{e^{-r/a_0}}{\sqrt{4\pi}} \right)^2 \frac{1}{|\vec{R} - \vec{r}|}$$

arising from the Coulomb interaction of the electron in a hydrogen atom with another charged particle at $\vec{R}$. The asymptotic result for $R \to \infty$ is $1/R$, and one might expect to calculate corrections as a power series in $a_0/R$ by expanding the denominator in powers of $r/R$. Show that all such corrections vanish. (Hint: consider the generating function for Legendre polynomials.) The power series is thus, trivially, convergent — but it does not give the right answer. Evaluate the integral exactly and show that there is a correction exponentially small in $a_0/R$.

**Exercise 4.2.** Consider the problem of evaluating the integral $F \equiv \int_A^B dx f(x)$, between specific endpoints $A$ and $B$, when nothing is known about the function $f(x)$ except for the first few terms of its two Taylor series, about $A$ and about $B$. The natural approximants are

$$F^{[n,m]}(\xi) \equiv \int_A^\xi dx f_A^{[n]}(x) + \int_\xi^B dx f_B^{[m]}(x),$$

where $f_A^{[n]}$ is the series about $A$, truncated after the $(x - A)^n$ term, and similarly for $f_B^{[n]}$. Clearly, these approximants depend on the extraneous variable $\xi$, though $F$ itself does not. Show that the PMS criterion fixes $\xi$ to be where the curves from the two endpoints cross, if they do.

Consider the specific example $f(x) = \sin x$ with $A = 0$, $B = \pi/2$. Note that the zeroth-order result $F^{[0,0]}(\xi) = \frac{\pi}{2} - \xi$ is monotonic in $\xi$. Find and plot the diagonal approximants $F^{[k,k]}(\xi)$ for $k = 0$ to 5. Apply the PMS criterion and compare and contrast with the results from a FAC criterion requiring either (i) no change from one order to the next, so that adjacent orders agree, or (ii) no change from the zeroth-order result. Verify the results in the table below and note the following: FAC(i) chooses $\xi$ to be 0 or $\pi/2$ and thus gives very poor results; FAC(ii) typically chooses a $\xi$ around 0.57, on the outskirts of the flat region; the centre of the flat region, the PMS choice, yields good results consistently.

|         | 1    | 2     | 3      | 4       | 5       |
|---------|------|-------|--------|---------|---------|
| PMS     | 1.07 | 0.996 | 0.9982 | 1.00005 | 1.00002 |
| FAC(i)  | 1.57 | 1.23  | 0.925  | 0.980   | 1.0045  |
| FAC(ii) | 1.57 | 0.996 | 0.9921 | 1.00015 | 1.0020  |

Consider other choices of $f(x)$. Note that when the function has a lot a structure between $A$ and $B$, the results will inevitably be poor, until the Taylor series contain enough terms to begin to describe the structure adequately. Beyond that point you should find the PMS criterion giving satisfactory results.

**Exercise 4.3.** Consider a quantum particle of unit mass that moves in the $x_1, x_2$ plane subject to a potential $\lambda x_1^2 x_2^2$. The Hamiltonian is

$$\tfrac{1}{2}p_1^2 + \tfrac{1}{2}p_2^2 + \lambda x_1^2 x_2^2.$$

By adding and subtracting a term $\tfrac{1}{2}\Omega(x_1^2 + x_2^2)$ use first-order perturbation theory and the PMS to find an approximation to the energy levels (labelled by $n_1, n_2$ quantum numbers).

**Exercise 4.4.** Consider the effect of an external electric field $\boldsymbol{E}$ on the $n = 2$ levels of hydrogen, ignoring spin and fine-structure effects. Standard first-order degenerate perturbation theory predicts a pattern where two levels are unaffected, while two other linear combinations of states are pushed up and down by an equal energy splitting $\pm\sigma\mathbb{R} \equiv \pm 3ea_0\,|\,\boldsymbol{E}\,|$, where $\mathbb{R}$ is the Rydberg constant, $a_0$ is the Bohr radius, and $e$ is the electron charge. Experimentally, however, the observed splitting is slightly asymmetric. That can be predicted by a laborious second-order calculation, or by an improved version of the first-order calculation where we write the Hamiltonian as $H_0(Z) + H_{\text{int}}(Z)$ with

$$H_0(Z) = \frac{p^2}{2m} - \frac{Ze^2}{r},$$

$$H_{\text{int}}(Z) = (Z-1)\frac{e^2}{r} + e\boldsymbol{E}.\boldsymbol{r}.$$

Calculate the matrix elements of $H_{\text{int}}(Z)$ between the four unperturbed states and diagonalize. Determine the optimal $Z$ for each eigenvalue. Show, using a secondary approximation $\sigma \ll 1$, that the splittings are $\sigma(1 - \sigma + \cdots)$ and $-\sigma(1 + \sigma + \cdots)$ times $\mathbb{R}$.

# Chapter 5

# Induced Convergence

## 5.1. Three Regions: Diverging, Flat, and Overdamped

In non-invariant approximations the optimized value of the extraneous parameter evolves from one order to the next — and that fact can be vital to the convergence of the results. For the anharmonic oscillator, the CK expansion is divergent for any fixed value of the extraneous parameter $\Omega$. However, as we shall see, the CK results converge if we use the optimized value of $\Omega$, which steadily increases with order. We call this phenomenon "induced convergence."

We begin by considering a very simple example which, while not actually illustrating "induced convergence," is nevertheless quite instructive about high orders in non-invariant approximations. Consider a simple harmonic oscillator of some given frequency $\omega$ and a calculation of its ground-state energy $E_0$ in a CK-like perturbation theory:

$$H_0 = \frac{1}{2}(p^2 + \Omega^2 x^2), \qquad H_{\text{int}} = -\frac{1}{2}(\Omega^2 - \omega^2)x^2. \qquad (5.1)$$

That is, we start from an oscillator of a different frequency $\Omega$, and treat the frequency difference as a perturbation. It is easy to see that the resulting perturbation series is the expansion of

$$E_0 = \frac{1}{2}\Omega\sqrt{1 - z} \qquad (5.2)$$

in powers of $z$, where

$$z \equiv 1 - \omega^2/\Omega^2, \qquad (5.3)$$

with finite-order approximations corresponding to successive truncations of the series

$$E_0 = \frac{1}{2}\Omega \left( 1 - \frac{1}{2}z - \frac{1}{8}z^2 - \frac{1}{16}z^3 - \frac{5}{128}z^4 + \cdots \right). \qquad (5.4)$$

Not surprisingly, the optimal choice of $\Omega$ in any order is $\Omega = \omega$,
and it always yields the exact result, $E_0 = \frac{1}{2}\omega$. Nevertheless, it is
instructive to plot the results, in $k$th order, as a function of $\Omega$; see
Fig. 5.1. As the order increases, one sees the curves becoming flatter
and flatter around $\Omega = \omega$, with the first to the $k$th derivatives all
vanishing there in $k$th order. The connection between the flatness and
the accuracy of the approximation is evident. Note that the results
diverge for $\Omega < \omega/\sqrt{2}$, reflecting the fact that the series has a radius
of convergence $|z| = 1$.

The figure illustrates the three regions that seem to be characteristic of non-invariant approximations. There is a "diverging region"
at small $\Omega$ where the results show wild oscillations from one order to
the next; a "flat region" around the optimal $\Omega$; and an "overdamped
region" at large $\Omega$. The extent of the flat region grows with the
order $k$. Usually the centre of the flat region also moves, but in

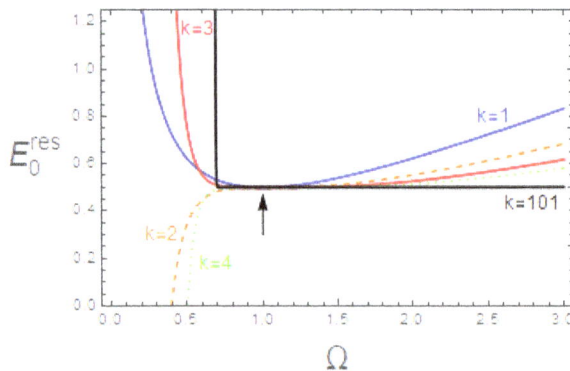

Fig. 5.1.   Results for $E_0$, as a function of the extraneous variable $\Omega$, in $k$th order
of a CK-like expansion for a simple harmonic oscillator of unit frequency ($\omega = 1$).
In every order the optimal $\Omega$ is 1, as indicated by the arrow. The curves become
flatter and flatter around $\Omega = 1$ as the order is increased.

this simple case it remains fixed at $\Omega = \omega$. In this example, the results ultimately converge at any $\Omega$, provided it exceeds $\omega/\sqrt{2}$. However, if we choose an $\Omega$ that is far too big — in the "overdamped region" — the convergence will be painfully slow (just as a strongly overdamped oscillator in classical mechanics takes a long time to return to equilibrium). Worse still, if we were to choose a too-small $\Omega$, in the "diverging region" $\Omega < \omega/\sqrt{2}$, then our initial bad result only becomes worse and worse as we calculate more and more orders!

In other contexts, the curves in the overdamped region may fall towards zero, rather than rising up to infinity, as they head away from the flat region. Also, the three regions may appear from right to left instead of from left to right; that occurs, for example, if we redefine the extraneous parameter from $\Omega$ to $1/\Omega$. The three regions show up in all the examples we study here.

## 5.2. High Orders in the CK Expansion for the Quartic Oscillator

The CK expansion is divergent for any fixed $\Omega$, just like ordinary perturbation theory for the anharmonic oscillator — where it is known that $k$th-order correction term behaves as $k!A^k k^B C(1 + O(1/k))\lambda^k$ with calculable numerical coefficients $A, B, C$. In $k$th order the CK result for $E_0$ can be written as a polynomial in $\lambda/\Omega^3$, which is the inverse of Caswell's variable $\beta$:

$$E_0^{[k]} = \Omega \left( A_{k,0} + A_{k,1} \left( \frac{\lambda}{\Omega^3} \right) + \cdots + A_{k,k} \left( \frac{\lambda}{\Omega^3} \right)^k \right). \qquad (5.5)$$

The first coefficient $A_{k,0}$ comes from just the $x^2$ term in $H_{\text{int}}$, so it is predictable from the simple example just discussed. The highest coefficient $A_{k,k}$ comes only from the $\lambda x^4$ term in $H_{\text{int}}$ and is the same as in ordinary perturbation theory. Caswell derives a recursion relation that allows the $A_{k,j}$ coefficients to be found efficiently.

> **This polynomial** can be seen as a truncated expansion in an effective dimensionless coupling constant $\lambda/\Omega^3$. However, it is not the perturbation series itself, since the terms are not successive contributions from matrix elements of $H_{\text{int}}$. Indeed, it is not successive truncations of a fixed series: the coefficients $A_{k,j}$ all

depend upon $k$. The series (5.5) displays quite good apparent convergence at the optimal $\Omega$ value.

For any fixed $\Omega$ the perturbation series itself diverges, but when $\Omega$ is chosen in each order according to the "minimal sensitivity" criterion — so that it gradually increases with order — one finds the successive results converging quite nicely. These points are illustrated below.

Figure 5.2 shows orders up to $k = 5$. One sees the three regions, with the flat region expanding, but also moving out to larger $\Omega$. It moves faster than it grows. Thus, for any fixed $\Omega$ one gets typical divergent-asymptotic-series behaviour. For example, with $\Omega/\lambda^{1/3} = 2.1$ the initial result is poor, but the results improve steadily for the next few orders, appearing to converge to a good answer — and then, quite abruptly, there are wilder and wilder oscillations from one order to the next; see Fig. 5.3. Choosing $\Omega$ a bit smaller would give better results at first order, but the divergent behaviour would set in almost immediately. (Choosing $\Omega$ too small puts one in the diverging region right away.) Choosing $\Omega$ to be larger one stays in the overdamped region for a longer time, with worse results at low orders and a more gradual approach towards settling down — though now to a more precise value — before the divergent behaviour finally sets in.

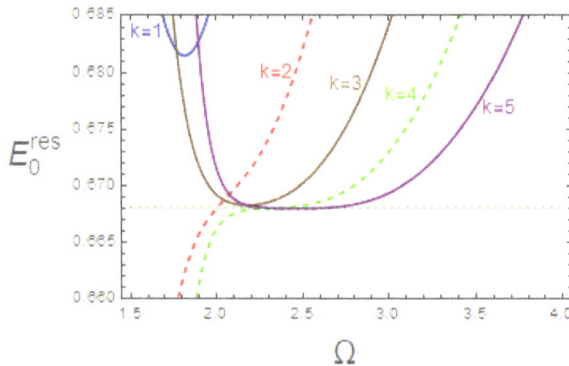

Fig. 5.2.   Results for $E_0$ to $k$th order in the CK expansion, as a function of the extraneous variable $\Omega$, in units of $\lambda^{1/3}$. The flat region grows, but it moves faster than it grows. For any fixed $\Omega$ the results diverge, but if we follow the flat region, finding the PMS optimal $\Omega$ in each order, the results converge nicely.

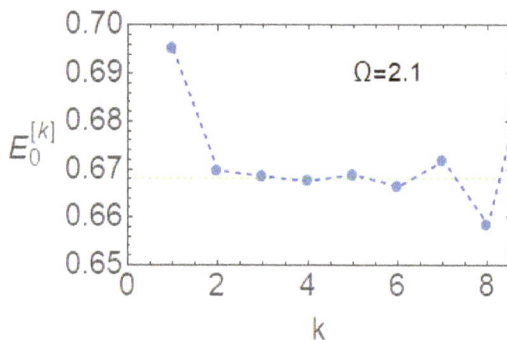

Fig. 5.3.   An illustration of the asymptotic-series behaviour at fixed $\Omega$. Results for $E_0$ at $k$th order for $k = 1$ to 8 at $\Omega = 2.1$.

With PMS optimization we ensure that a suitable $\Omega$ is used at *each* order, starting with $\Omega/\lambda^{1/3} = 6^{1/3} = 1.82$ at $k = 1$ and steadily increasing with $k$, so that we "surf" along with the flat region as it moves outwards. In fact, we want to "surf the scary edge" of the flat region, close to the boundary with the diverging region. This point is illustrated in Fig. 5.4, which shows results for 23rd order. Here, when we examine the flat region in detail, we find three stationary points. All give a good approximation, but by far the best approximation comes from the one closest to the diverging region — which is also the "flattest" in that it has the smallest second derivative. As $k$ is increased, each individual stationary point moves off to the right and new stationary points are born at the "scary edge." When being born these new stationary points may appear in embryo as points of inflexion.

For the proof of convergence we refer the reader to references in the bibliography. The proof applies also to a $\phi^4$ field theory in $1 + 1$ dimensions and shows that the convergence is exponentially fast.

The shrinkage of the effective coupling $\lambda/\Omega^3$ with increasing order is readily understandable intuitively through an extension of the argument in Sec. 4.5 and Fig. 4.4. Quantum mechanical perturbation theory for $E_n$ at $k$th order involves excursions to various intermediate states $|n_i\rangle$, producing a product of matrix elements

$$\langle n | H_{\text{int}} | n_2 \rangle \langle n_2 | \cdots | n_k \rangle \langle n_k | H_{\text{int}} | n \rangle, \tag{5.6}$$

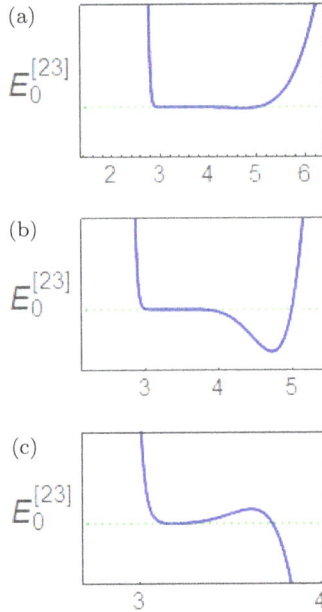

Fig. 5.4.   Results for $E_0$ at 23rd order in the CK expansion, as a function of $\Omega$: (a) shows all three regions; diverging, flat, and overdamped, (b) shows the flat region on a fine scale, and (c) shows a very fine-scale close up of the front of the flat region. The vertical ranges are, respectively, (a) 0.667 to 0.670, (b) 0.66796 to 0.668025, (c) 0.6679861 to 0.6679865. The exact result, shown as a dotted line, is 0.66798626. Of the three stationary points in the flat region the one closest to the onset of the divergent region is the most accurate.

each with its corresponding energy denominator $E_n - E_{n'}$. In very high orders most of these matrix elements involve high-lying eigenstates — and for such states, if the unperturbed states and energies are to mimic the actual ones, the appropriate $\Omega$ is large. Thus, we can expect the optimal $\Omega$ to steadily increase with order.

The same intuitive argument applies to the QFT case (provided we consider only perturbatively calculable, infrared-safe quantities). If we were to use "old-fashioned perturbation theory" then, again, the high orders would involve the physics of short-lived, high-energy intermediate states. In covariant Feynman perturbation theory the corresponding statement is that the high orders will involve highly virtual particles. In QCD, of course, the effective couplant shrinks only logarithmically with energy scale, rather than the $\lambda/\Omega^3$ behaviour of the oscillator problem.

## 5.3. A Toy Model Framework

For the remainder of this chapter we consider a toy model. It is a bit artificial, but has the virtue that the induced convergence property can be proved relatively easily.

Given a series

$$\mathcal{R} = a_0(1 + r_1(0)\, a_0 + r_2(0)\, a_0^2 + \cdots), \tag{5.7}$$

let us make the substitution

$$a \equiv a(\tau) = \frac{a_0}{1 + \tau a_0}, \tag{5.8}$$

where $\tau$ is some real-valued parameter, and consider the resulting re-expansions

$$\mathcal{R} = a(1 + r_1 a + r_2 a^2 + \cdots). \tag{5.9}$$

The coefficients are easily determined to be

$$r_j \equiv r_j(\tau) = \sum_{i=0}^{j} \binom{j}{i} \tau^i r_{j-i}(0), \tag{5.10}$$

where

$$\binom{j}{i} = \frac{j!}{i!(j-i)!}, \tag{5.11}$$

are the binomial coefficients. This simple mathematical system mimics, to some extent, the RS dependence problem. Both the expansion parameter $a$ and the coefficients $r_j$ in Eq. (5.9) depend on the extraneous variable $\tau$, which plays the role of the RS. Obviously, $\mathcal{R}$ itself does not depend on $\tau$, and the $\tau$ dependences of $a$ and the $r_j$ must cancel in Eq. (5.9). However, this cancellation is spoiled if the series is truncated.

For the analogy with RS dependence to hold good it is important that all the expansion parameters should be on an equal footing. Equation (5.8) is the simplest example of a substitution which achieves this. The symmetry between any two expansion parameters is made manifest by the relation

$$\frac{1}{a} - \frac{1}{a'} = \tau - \tau'. \tag{5.12}$$

In contrast, a substitution such as $a = a_0(1 + \tau a_0)$ would not have been suitable, since it would have given $a_0$ a unique status. Equation (5.8) is particularly appropriate since it directly recalls the (leading-order) QCD running coupling constant formula, if we identify $\tau$ as proportional to the logarithm of the renormalization scale. Indeed, one has the "$\beta$-function equation"

$$\frac{da}{d\tau} = -a^2. \tag{5.13}$$

The analogy with renormalized perturbation theory is not perfect. Here the multiplicity of expansion parameters, the "extraneous" variable, the "$\beta$ function," and the resulting ambiguity of finite-order results have all been introduced artificially. Moreover, only a single extraneous variable has been introduced here, whereas the real problem involves $(n - 1)$ RS parameters at $n$th order. Nevertheless, this oversimplified model can be very instructive.

Differentiating Eq. (5.9) and using Eq. (5.13) leads to

$$\frac{d\mathcal{R}}{d\tau} = (-a^2)(1 + 2r_1 a + 3r_2 a^2 + \cdots) + (\dot{r}_1 a^2 + \dot{r}_2 a^3 + \dot{r}_3 a^4 + \cdots). \tag{5.14}$$

The cancellations required for this to vanish yield

$$\dot{r}_j \equiv \frac{dr_j}{d\tau} = j\, r_{j-1}, \tag{5.15}$$

which determine the $\tau$ dependence of the coefficients. Integration of these equations leads, of course, to Eq. (5.10), previously obtained by direct substitution.

The $n$th-order approximant, defined as the truncated series

$$\mathcal{R}^{(n)} \equiv a(1 + r_1 a + \cdots + r_{n-1}a^{n-1}), \tag{5.16}$$

is now a known, well-defined function of $\tau$, as soon as we know the values of $a$ and $r_1, \ldots, r_{n-1}$ at some fixed value of $\tau$, such as $\tau = 0$.

It is not often necessary to have an explicit formula for $\mathcal{R}^{(n)}(\tau)$, but it is sometimes helpful. One may write $\mathcal{R}^{(n)}(\tau)$ as the rational function

$$\mathcal{R}^{(n)}(\tau) = \frac{1}{(a_0^{-1} + \tau)^n} \sum_{j=1}^{n} \tau^{n-j} a_0^{-j} \binom{n}{j} \mathcal{R}^{(j)}(0). \qquad (5.17)$$

This formula is most easily proved by induction: a direct proof requires a remarkable identity between binomial coefficients.

The derivative of $\mathcal{R}^{(n)}$ is the residuum of the cancellations in Eq. (5.14):

$$\frac{d\mathcal{R}^{(n)}}{d\tau} = -n r_{n-1} a^{n+1}. \qquad (5.18)$$

The "optimum" value of $\tau$ at $n$th order, according to the PMS criterion, is where the right-hand side vanishes, which leads to the "optimization condition"

$$r_{n-1}(\tau = \bar{\tau}_n) = 0. \qquad (5.19)$$

(In this instance, the PMS condition is equivalent to a FAC-type condition that the last calculated coefficient is made to vanish.) After solving this equation for $\bar{\tau}_n$, the "optimized" approximant $\mathcal{R}_{\text{opt}}^{(n)}$ can be found by evaluating $\mathcal{R}^{(n)}(\tau)$ at the point $\tau = \bar{\tau}_n$.

> **If the PMS condition**, Eq. (5.19), has no real solution — a situation that will arise in odd orders in the following example — one has two reasonable choices, as discussed in Sec. 4.6. One can consider solutions of $d^2\mathcal{R}^{(n)}/d\tau^2 = 0$, or one can use a complex solution of Eq. (5.19), finally discarding the small imaginary part of the resulting $\mathcal{R}^{(n)}$. In the real RS-optimization problem, with multiple RS variables, one always finds a solution to the usual PMS condition, so we need not dwell on this issue.

Clearly, one can investigate many specific examples within this framework, simply by beginning with a different initial series in Eq. (5.7). It is a simple exercise to show that if the initial series is geometric, then the optimized result in second order, and all subsequent orders, is exact. This is true irrespective of whether the initial series was within its radius of convergence or not. In the next section we consider a more challenging example in detail.

## 5.4. Alternating Factorial Series

Consider the alternating factorial series

$$\mathcal{R} = a_0(1 - 1!a_0 + 2!a_0^2 - 3!a_0^3 + \cdots), \qquad (5.20)$$

previously discussed after Eq. (4.2). In the transformed series, Eq. (5.9), the coefficients are given by

$$r_j = (-1)^j \, j! \sum_{i=0}^{j} \frac{(-\tau)^i}{i!} = (-1)^j \, j! \, \mathbb{T}_j[e^{-\tau}], \qquad (5.21)$$

where the notation $\mathbb{T}_n[F(a)]$ means "truncate the series for $F(a) = F_0 + F_1 a + \cdots$ immediately after the $a^n$ term" (i.e., $\mathbb{T}_n[F(a)] \equiv F_0 + F_1 a + \cdots + F_n a^n$.). If $\tau$ is fixed then the coefficients will eventually behave as $(-1)^j j! e^{-\tau}$ for sufficiently large $j$, with $e^{-\tau}$ being a constant. Therefore, the series remains divergent for all fixed values of $\tau$.

However, *if we "optimize" the choice of $\tau$ at each order then the resulting sequence of approximations is convergent.* The proof will be given in the next section. First, we discuss the optimization procedure in more detail and present a numerical example.

In this case the optimization condition in $n$th order, Eq. (5.19), becomes simply

$$\mathbb{T}_{n-1}[e^{-\bar{\tau}_n}] = 0. \qquad (5.22)$$

For even values of $n$ this equation has a single real root, which grows approximately linearly with $n$ (see Eq. (5.26) below).

**For odd** $n$ there is no real root. As mentioned earlier, one may either use a complex root, or seek to minimize the slope, so that $d^2\mathcal{R}^{(n)}/d\tau^2 = 0$, which leads to

$$\left(a_0^{-1} + \bar{\tau}_n\right) \mathbb{T}_{n-2}[e^{-\bar{\tau}_n}] + (n+1)\mathbb{T}_{n-1}[e^{-\bar{\tau}_n}] = 0.$$

This equation has two real roots, but the larger corresponds to a maximum of the slope, and hence is not relevant. The appropriate root lies just beyond the $\bar{\tau}$ of the previous even order.

Having solved for $\bar{\tau}_n$ one can immediately evaluate the optimized values of $a$, $r_j$ and $\mathcal{R}^{(n)}$ from Eqs. (5.8), (5.16), (5.21). (Equivalently, one could substitute in Eq. (5.17).) Some numerical results for the case $a_0 = 0.25$ are shown in Table 5.1 and Fig. 5.5.

The naïve results, corresponding to the partial sums of Eq. (5.20), show some initial apparent convergence, but soon develop violent fluctuations. The optimized results, by contrast, converge in a steady fashion. This comes about because $\bar{\tau}_n$ grows with $n$ at just the right rate. The resulting decrease in the effective expansion parameter $\bar{a}_n$ counterbalances the potential growth of the coefficients. The radius of the original series is zero, but asymptotically one is expanding in powers of a vanishing parameter. Note that it is convergence of a *sequence* — the sequence of optimized approximants — that really matters, not the convergence of any *series*.

Table 5.1. Approximations to the alternating factorial series, for $a_0 = 0.25$. The Borel sum has the value 0.20634565.

| Order | Naive | $\bar{\tau}$ | $\bar{a}$ | Optimum |
|---|---|---|---|---|
| 1 | 0.250 | – | – | – |
| 2 | 0.188 | 1.00000 | 0.200 | 0.2000000 |
| 3 | 0.219 | 1.43845 | 0.184 | 0.2061121 |
| 4 | 0.195 | 1.59607 | 0.179 | 0.2054645 |
| 5 | 0.219 | 2.00000 | 0.167 | 0.2062757 |
| 6 | 0.189 | 2.18061 | 0.162 | 0.2061192 |
| 7 | 0.233 | 2.59875 | 0.152 | 0.2063073 |
| 8 | 0.156 | 2.75900 | 0.148 | 0.2062634 |
| 9 | 0.310 | 3.19736 | 0.139 | 0.2063235 |
| 10 | −0.036 | 3.33355 | 0.136 | 0.2063083 |
| 11 | 0.829 | 3.79230 | 0.128 | 0.2063320 |
| 12 | −1.550 | 3.90545 | 0.126 | 0.2063259 |
| 13 | 5.588 | 4.38352 | 0.119 | 0.2063367 |
| 14 | −17.610 | 4.47541 | 0.118 | 0.2063340 |
| 15 | 63.581 | 4.97148 | 0.111 | 0.2063394 |

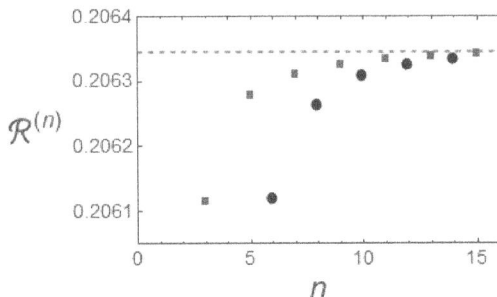

Fig. 5.5. Optimized results for the alternating factorial series example. Even orders shown by dots, odd orders by squares. The Borel sum is indicated by the dashed line.

Having said that, it is nevertheless interesting to see how the terms of the (truncated) series behave numerically at the "optimized" value of $\tau$. A typical example is 8th order, which looks like

$$\mathcal{R}^{(8)}(\bar{\tau}_8) = 0.148(1 + 0.260 + 0.090 + 0.028 + 0.011$$

$$+ 0.003 + 0.002 + 0). \tag{5.23}$$

Thus, the "apparent convergence" of the series in the "optimized" scheme is perfectly satisfactory. The residual error one might reasonably guess from this, say, about $0.148 \times 0.001$, is a reasonable indicator of the actual error (which is, in fact, about half this size). Note that our estimate here used the trend of the last few terms, not just the last term alone — because it, of course, was made to vanish exactly by the optimization condition. (One may also make an error estimate based on the sequence of optimized results in Table 5.1, with reasonable results.)

It is also instructive to plot some graphs of $\mathcal{R}^{(n)}(\tau)$ against $\tau$ for various values of $n$. This has been done in Fig. 5.6, by utilizing Eq. (5.17). As before, one can describe each curve in terms of three main regions: (i) a "diverging region" at small $\tau$ where the function tends rapidly to $+\infty$ or $-\infty$, depending on whether $n$ is odd or even, (ii) a "flat region" around $\tau = \bar{\tau}_n$, and (iii) an "overdamped region" at large $\tau$ in which $\mathcal{R}^{(n)}(\tau)$ is small. The fuzzy boundaries between the regions move steadily to the right as $n$ increases. They do so at slightly different rates, so that the width of the flat region slowly grows.

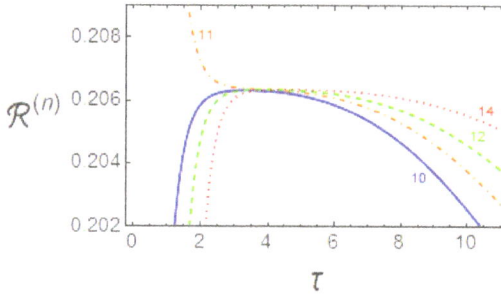

Fig. 5.6. Approximants $\mathcal{R}^{(n)}(\tau)$, $n = 10, 11, 12, 14$, for the alternating factorial series example, shown as functions of the extraneous variable $\tau$.

If we sit at some fixed, reasonably large $\tau$, we see each of these regions in turn, as $n$ increases. Low-order results are too small, but they grow steadily and begin to settle down to an almost constant value. Then, quite suddenly, we are overtaken by the diverging region, and the results start to oscillate increasingly violently from one order to the next. The view from the "comoving frame" of the optimized $\tau$ is quite different. One sees the approximation converging smoothly, and becoming flat over an increasing range of $\tau$. From this vantage point the approximation is not only becoming steadily more accurate, it is also successfully mimicking the $\tau$-independence property of the exact result.

## 5.5. Proof of Convergence

This section outlines the proof that the sequence of optimized approximants does indeed converge. One first needs to know the behaviour of $\bar{\tau}_n$ as $n \to \infty$. For this purpose one needs the integral representation

$$\mathbb{T}_{n-1}[e^{-\tau}] = e^{-\tau}\left(1 + \frac{(-1)^{n-1}}{\Gamma(n)}\int_0^\tau dt\, t^{n-1}e^t\right), \qquad (5.24)$$

obtained by repeated integration by parts. In the case that $n$ is even, the optimization condition $\mathbb{T}_{n-1}[e^{-\tau_n}] = 0$ reduces to

$$\int_0^{\bar{\tau}_n} dt\, t^{n-1}e^t = \Gamma(n). \qquad (5.25)$$

One can now consider this as an abstract problem with $n$ ranging over all real values. One can take the difference of this equation and the corresponding equation with $n \to n+1$ and then put bounds on the remaining integral term. Then, using the asymptotic behaviour of the $\Gamma$ function, one finds that

$$\bar{\tau}_n \sim \chi n + \xi \ln n + O(1),  \tag{5.26}$$

where $\chi \approx 0.278$ is the solution to

$$\ln \chi + \chi + 1 = 0,  \tag{5.27}$$

and $\xi = \frac{1}{2}\chi/(1+\chi) \approx 0.109$. For odd $n$ an analysis of the $d^2\mathcal{R}^{(n)}/d\tau^2 = 0$ equation leads to the same result, except that the coefficient of the $\ln n$ term in Eq. (5.26) is altered.

The alternating factorial series, Eq. (5.20), is a classic example of a Borel-summable divergent series. The sum of the series, in the sense of Borel, is given by the integral

$$\mathcal{R}_B \equiv \int_0^\infty du\, e^{-u/a_0} \frac{1}{1+u}.  \tag{5.28}$$

Heuristically, the series is generated by expanding $1/(1+u)$ as $1 - u + u^2 - u^3 + \cdots$ and then integrating term by term. Of course, the expansion is only valid for $|u| < 1$ whereas the integral involves $0 < u < \infty$. The difficulty is exposed by using the formula for the sum of a truncated geometric series:

$$\mathbb{T}_{n-1}\left[\frac{1}{1+u}\right] = \frac{1-(-u)^n}{1+u}.  \tag{5.29}$$

Thus, truncations of Eq. (5.20) correspond to

$$\mathcal{R}^{(n)}(0) = \int_0^\infty du\, e^{-u/a_0}\left(\frac{1-(-u)^n}{1+u}\right),  \tag{5.30}$$

in which the $(-u)^n$ term is far from negligible as $n \to \infty$, and is the cause of the wild oscillations. Recalling that $1/a_0 = 1/a - \tau$, one can write down a similar integral representation of the $n$th-order approximant at a general value of $\tau$:

$$\mathcal{R}^{(n)}(\tau) = \int_0^\infty du\, e^{-u/a}\, \mathbb{T}_{n-1}\left[e^{u\tau}\frac{1}{1+u}\right]. \tag{5.31}$$

The following algebraic manipulations:

$$\mathbb{T}_{n-1}\left[e^{u\tau}\frac{1}{1+u}\right] = \mathbb{T}_{n-1}\left[\frac{1}{1+u}\right] + \frac{u\tau}{1!}\mathbb{T}_{n-2}\left[\frac{1}{1+u}\right]$$

$$+\cdots+\frac{(u\tau)^{n-1}}{(n-1)!}\mathbb{T}_0\left[\frac{1}{1+u}\right] = \left(\frac{1-(-u)^n}{1+u}\right)$$

$$+\frac{u\tau}{1!}\left(\frac{1-(-u)^{n-1}}{1+u}\right)+\cdots+\frac{(u\tau)^{n-1}}{(n-1)!}\left(\frac{1-(-u)^1}{1+u}\right)$$

$$= \frac{1}{1+u}\left(1+\frac{u\tau}{1!}+\cdots+\frac{(u\tau)^{n-1}}{(n-1)!}\right) - \frac{(-u)^n}{1+u}$$

$$\times\left(1+\frac{-\tau}{1!}+\cdots+\frac{(-\tau)^{n-1}}{(n-1)!}\right)$$

$$= \frac{1}{1+u}\left(\mathbb{T}_{n-1}[e^{u\tau}] - (-u)^n\mathbb{T}_{n-1}[e^{-\tau}]\right), \tag{5.32}$$

then lead to

$$\mathcal{R}^{(n)}(\tau) = \int_0^\infty du\, \frac{e^{-u/a}}{1+u}\left(\mathbb{T}_{n-1}[e^{u\tau}] - (-u)^n\mathbb{T}_{n-1}[e^{-\tau}]\right). \tag{5.33}$$

The optimization condition chooses $\tau$ such that $\mathbb{T}_{n-1}[e^{-\tau}] = 0$, and so *it eliminates the troublesome* $(-u)^n$ *term*. (Let us restrict the discussion to even values of $n$ for the moment.) If $\mathbb{T}_{n-1}[e^{u\tau}]$ could be replaced by $e^{u\tau}$ as $n \to \infty$, then one would have established that the sequence of optimized approximants converges to $\mathcal{R}_{\mathrm{B}}$.

However, that replacement is not valid when $\tau$ is $\bar{\tau}_n$, which is linearly increasing with $n$. Each term in the difference $(e^{u\bar{\tau}_n} - \mathbb{T}_{n-1}[e^{u\bar{\tau}_n}])$ generates only an $O(1/n)$ correction, but the sum of those corrections is finite. To show that fact one can use the representation equation (5.24) to rewrite Eq. (5.33), for $\tau = \bar{\tau}_n$, as

$$\mathcal{R}_{\mathrm{opt}}^{(n)} = \mathcal{R}_{\mathrm{B}} - \int_0^\infty du\, \frac{e^{-u/a_0}}{1+u}\frac{\gamma(n, u\bar{\tau}_n)}{\Gamma(n)}, \tag{5.34}$$

where $\gamma(n, z)$ is the incomplete $\Gamma$ function

$$\gamma(n, z) \equiv \int_0^z dv \, v^{n-1} e^{-v}, \tag{5.35}$$

and $\Gamma(n)$ is, of course, $\gamma(n, \infty)$. One requires the large-$n$ behaviour of $\gamma(n, z(n))$ when $z$ behaves as $u\chi n + O(\ln n)$. Applying Laplace's method, one finds the result (see references in the Bibliography)

$$\lim_{n\to\infty} \left( \frac{\gamma(n, \chi un)}{\Gamma(n)} \right) = \begin{cases} 0, & u\chi < 1, \\ 1, & u\chi > 1. \end{cases} \tag{5.36}$$

Laplace's method can also be used to show that the result is unchanged when the $\xi \ln n$ term in $\bar{\tau}_n$, Eq. (5.26), is included. (This indicates, incidentally, that the flat region grows at least as fast as $\ln n$.) From the above equation, we see that the second term in Eq. (5.34) receives contributions only for $u > 1/\chi \approx 3.591$, and its integrand has exactly the same form as that of $\mathcal{R}_B$. Thus, the limit of the sequence of optimized approximants is *not* the same as the Borel sum: Instead,

$$\lim_{n\to\infty} \mathcal{R}_{\text{opt}}^{(n)} = \int_0^{1/\chi} du \, e^{-u/a_0} \frac{1}{1+u}. \tag{5.37}$$

The difference from $\mathcal{R}_B$ lies only in the upper limit of the integral — something that does not affect the series expansion. The difference is exponentially small, $O(a_0 e^{-1/(\chi a_0)})$ for small $a_0$. In the numerical example of Table 5.1 with $a_0 = 0.25$, the difference is $3 \times 10^{-8}$. (This difference is far too small to be visible in the figures earlier.)

> **The above proof** was actually restricted to the case $n$ even. When $n$ is odd we also need to consider the $\mathbb{T}_{n-1}[e^{-\tau}]$ term in Eq. (5.33). One can show, by an extension of Laplace's technique, that this term gives a contribution that is suppressed by a $1/n$ factor, so that odd orders converge to the same limit as even orders. (The $1/n$ factor was unfortunately mislaid in Ref. [10], which erroneously stated that odd orders converged to a different limit.)

We caution that one should not necessarily assume that the Borel sum is "right" and the optimized limit is slightly "wrong" — nor *vice versa*. As pointed out in Sec. 4.1 that issue can only be meaningfully addressed given a specific context in which the original divergent series arose. One of the most unsatisfactory features of our "toy model" framework is just that it has no such context, since we began with "Given a series ...". Also, here one expansion parameter is actually a bit special, because the expansion in $a_0$ has a particularly simple and regular form, compared to general $a$ — whereas in the real problem all $a$'s are fundamentally on an equal footing, with the counterpart to $a_0$ being the ill-defined bare coupling constant.

## 5.6. Conclusions

This chapter has only scratched the surface of a large subject that raises difficult mathematical challenges. The conditions under which induced convergence will occur, and whether it works in the context of OPT applied to QCD remain unanswered questions.

**Exercise 5.1.** Prove the claim at end of Sec. 5.3 that if the initial series is a geometric series, then the optimized result in second order, and all subsequent orders, is exact — irrespective of whether the initial series is within its radius of convergence or not.

**Exercise 5.2.** Prove Eq. (5.17) by induction.

**Exercise 5.3.** Prove Eq. (5.26) for the high-$n$ behaviour of $\bar{\tau}_n$.

**Exercise 5.4.** Find how Fig. 5.6 would appear for very large $n$, if $\mathcal{R}^{(n)}$ is plotted as a function of $t \equiv \tau/n$. By considering the second term of Eq. (5.33), show that the boundary between the divergent and flat regions is at $t = \chi$, so that the optimized result "surfs the scary edge."

# Part II

# Optimized Perturbation Theory

## Chapter 6

# Preliminaries: RG Invariance, int-$\beta$ Equation, $\tilde{\Lambda}$ Definition, and CG Relation

## 6.1. Physical Quantities and Their RG Invariance

QFT involves many theoretical objects — Green's functions, for instance — that are not physically measurable and are not RG invariant. While these are very important, they are not directly relevant to our concern here. Ultimately, it is only the theory's predictions for physical, experimentally measurable quantities that will matter. For our purposes then, Green's functions, etc., are merely intermediate steps in calculating physical quantities and we shall focus our discussion on the physical quantities themselves.

Some physical quantities, such as hadron masses, are inherently inaccessible to perturbation theory. Others have a factorized form where one factor is non-perturbative; such quantities will be discussed in Chapter 12. (Other physical quantities, such as the QCD pressure at high temperature, have a still more complex structure.) Here we focus on quantities with a normal perturbation series. These have the form $A_1 \mathcal{R} + A_0$ with a leading-order coefficient $A_1$ and, sometimes, a zeroth-order term $A_0$. The coefficients $A_0$ and $A_1$ (which carry the dimensions of mass to the appropriate power) are RS invariant, so we may focus on dimensionless physical quantities

$\mathcal{R}$ of the form:

$$\mathcal{R} = a^{\mathrm{P}}(1 + r_1 a + r_2 a^2 + \cdots), \qquad (6.1)$$

where $a$ is the renormalized couplant $g_R^2/(4\pi^2)$. The power $\mathrm{P}$ is typically 1 or 2 or 3, but may be left general. At the level of formal power series there would be no loss of generality in taking $\mathrm{P} = 1$, since one could always replace $\mathcal{R}$ with another physical quantity $\mathcal{R}^{1/\mathrm{P}}$. However, that trick would not be natural when $\mathcal{R}$ is to be truncated at finite order. In some places, to simplify the presentation, we will first discuss the case $\mathrm{P} = 1$, before generalizing to any $\mathrm{P}$.

Often $\mathcal{R}$ is not a single quantity but is a function of several experimentally defined parameters. We may always single out one parameter with the dimensions of mass that we will call the "physical energy scale" and denote by "$Q$." All the other experimental parameters can be taken to be dimensionless: if they were not originally we may scale them with an appropriate power of $Q$ to make them so. We will need "$Q$" only in order to discuss which quantities are, and which are not, $Q$ dependent. Thus, we may leave the specific definition of $Q$ in any particular case to the reader's choice, without creating any ambiguity.

The dependence of $\mathcal{R}$ on $Q$ is of great interest. When the original Lagrangian of the theory contains no parameter with dimensions of mass, but only a dimensionless bare coupling constant, $g_B$, then Dimensional Analysis would seem to imply that the dimensionless $\mathcal{R}$ must be independent of $Q$. As discussed in Chapter 2, that conclusion is false because of the need for renormalization — which necessarily introduces a renormalization scale, $\mu$. A calculation of the coefficient $r_1$ reveals that it depends logarithmically on the *ratio* of $Q$ to $\mu$ (and $r_2$ involves a log-squared, etc.). Although the $\mu$ dependence ultimately cancels in $\mathcal{R}$, the $Q$ dependence survives. Specification of a boundary condition for the $\beta$-function equation (see Sec. 6.3 below) introduces a mass parameter $\tilde{\Lambda}$, and $\mathcal{R}$ is ultimately a function of the ratio $Q/\tilde{\Lambda}$. The phenomenon that a theory, seemingly involving only a dimensionless constant $g_B$, ends up specified by a characteristic-scale parameter $\tilde{\Lambda}$ is known as "Dimensional Transmutation."

The key property of a *physical* quantity is that of "Renormalization Group" (RG) invariance, which means that a physical quantity is independent of the renormalization scheme (RS); that is, of the precise definition of the renormalized couplant, $a$. Heuristically, this property might be expressed symbolically as

$$0 = \frac{d\mathcal{R}}{d(RS)} = \frac{\partial \mathcal{R}}{\partial(RS)}\bigg|_a + \frac{da}{d(RS)}\frac{\partial \mathcal{R}}{\partial a}, \qquad (6.2)$$

where the total RS dependence is separated into two sources: (i) RS dependence from the series coefficients, $r_i$ (differentiation pretending that $a$ is constant), and (ii) RS dependence from the couplant, $a$.

**A note on notation:** The partial derivative $\frac{\partial \mathcal{R}}{\partial a}$ is the derivative regarding the coefficients $r_i$ as constant, and so is just the ordinary derivative of $\mathcal{R}$'s power series. There is a reason we do not write it as $d\mathcal{R}/da$; see Sec. 8.3.

A particular case of Eq. (6.2) is the familiar RG equation expressing the renormalization-scale independence of $\mathcal{R}$:

$$\left(\mu\frac{\partial}{\partial\mu}\bigg|_a + \beta(a)\frac{\partial}{\partial a}\right)\mathcal{R} = 0, \qquad (6.3)$$

which involves the famous "$\beta$ function," defined as

$$\beta(a) \equiv \mu\frac{da}{d\mu}. \qquad (6.4)$$

It is important to stress that there is *not* a unique $\beta$ function characteristic of a given theory. The $\beta$ function, like $a$, is an RS-dependent object. We shall write the $\beta$ function, in some arbitrary RS, in the form

$$\beta(a) = -ba^2(1 + ca + c_2a^2 + \cdots). \qquad (6.5)$$

As shown in the next section, the coefficients $b$ and $c$ are invariants, but all the higher coefficients $c_2, c_3, \ldots$ depend on RS.

**Note** that an explicit minus sign is included in front of the expansion of $\beta$ in Eq. (6.5) so that $b$ is a positive number in the most interesting ("asymptotically free") QFTs. However, all our discussion applies equally to theories with negative $b$, with the understanding that then the infrared and ultraviolet limits exchange roles. (For the special case of theories with $b = 0$ see Exercise 6.1.) The reason for factoring out $b$ in $\beta(a)$'s expansion is that it is then absent, or scales out of, almost all subsequent equations.

What exactly is the "renormalization scale" $\mu$? In defining a particular RS many arbitrary choices are made that, explicitly or implicitly, introduce many arbitrary parameters. One of these parameters necessarily has the dimensions of mass and may be singled out as "$\mu$;" any other parameters involved can be taken as dimensionless since otherwise we could always scale them with the appropriate power of $\mu$. It is sometimes helpful to use the terminology that a "renormalization prescription" (RP) is a set of renormalization conventions that fixes all the RS choices, including defining the "meaning" of $\mu$, but does not fix the *value* of $\mu$. (The $\overline{\text{MS}}$ scheme is really a "prescription" in this terminology.) Often "RP" and "RS" can be used interchangeably, but sometimes the distinction is useful to make.

The precise definition of $\mu$ in any specific RP, and the particular choice of $Q$ for any specific physical quantity, may be left to the whim of the reader. The results of optimized perturbation theory will be the same, in content, whatever choices are made.

## 6.2. RP Invariance of $b$ and $c$

A key fact, first appreciated by 't Hooft, is that the first two coefficients, $b$ and $c$, of the $\beta$ function are RP invariant, while the higher coefficients are not. The proof is straightforward: Two RP's (that is, two schemes with the same value of $\mu$) are related by a general transformation

$$a' = a(1 + v_1 a + v_2 a^2 + \cdots),  \tag{6.6}$$

where the $v_i$ coefficients cannot depend on $\mu$, for dimensional reasons. In the primed RP the $\beta$ function:

$$\beta'(a') \equiv \mu \frac{da'}{d\mu} = -b' a'^2 (1 + c' a' + c_2' a'^2 + \cdots)  \tag{6.7}$$

will not be the same as the original $\beta$ function. The two are related by

$$\beta'(a') \equiv \mu \frac{da'}{d\mu} = \frac{da'}{da} \mu \frac{da}{d\mu} = \frac{da'}{da} \beta(a).  \tag{6.8}$$

With $da'/da$ evaluated from Eq. (6.6) this gives

$$\beta'(a') = (1 + 2v_1 a + 3v_2 a^2 + \cdots)\beta(a). \qquad (6.9)$$

Expanding the right-hand side then leads to

$$(1 + 2v_1 a + 3v_2 a^2 + \cdots)(-ba^2)(1 + ca + c_2 a^2 + \cdots)$$
$$= -ba^2(1 + (2v_1 + c)a + (3v_2 + 2v_1 c + c_2)a^2 + \cdots), \qquad (6.10)$$

while eliminating $a'$ in favour of $a$ in the left-hand side gives

$$-b'\left(a(1 + v_1 a + v_2 a^2 + \cdots)\right)^2 (1 + c'a(1 + v_1 a + \cdots) + c_2' a^2 + \cdots)$$
$$= -b'a^2(1 + (2v_1 + c')a + (2v_2 + v_1^2 + 3v_1 c' + c_2')a^2 + \cdots). \qquad (6.11)$$

Equating coefficients of $a^2$ and $a^3$ shows that

$$b' = b \quad \text{and} \quad c' = c. \qquad (6.12)$$

Note that the higher coefficients are not invariant; for instance $c_2' = c_2 + v_2 - v_1^2 - v_1 c$, as one can see from equating coefficients of $a^4$ above.

**In gauge theories** some RP's are said to be "gauge dependent." This just means that the RP is not fully defined until the gauge choice is specified. As mentioned in Appendix 2.C, in such RP's the $\beta$ function should be defined with the $\mu \partial a/\partial \mu$ derivative taken at constant *renormalized* gauge parameter, $\xi$, not at constant bare $\xi_B$. With the other definition the second coefficient of the $\beta$ function would not be invariant because the renormalized $\xi$ would then be a source of implicit $\mu$ dependence in the $v_i$ coefficients. Note also that there is no need for a $\frac{d\xi}{d(RS)} \frac{\partial R}{\partial \xi}$ term in the RG equation, Eq. (6.2), because $\frac{\partial R}{\partial \xi} = 0$, since physical quantities are gauge invariant.

In any specific renormalizable QFT the invariants $b, c$ can be calculated and we regard them here as givens. For instance, in Quantum Chromodynamics (QCD) with $n_f$ flavours of massless quarks one has

$$b = \frac{33 - 2n_f}{6}, \quad c = \frac{153 - 19n_f}{2(33 - 2n_f)}. \qquad (6.13)$$

## 6.3. Integrated $\beta$-Function Equation; Definition of $\tilde{\Lambda}$

The $\beta$-function equation

$$\mu\frac{da}{d\mu} = \beta(a), \qquad (6.14)$$

integrates immediately to

$$\ln\mu = \int\frac{da}{\beta(a)} + \text{const.} \qquad (6.15)$$

However, specifying the constant of integration is a slightly tricky task and requires a careful discussion. The first step is to write the constant of integration as $\ln\tilde{\Lambda} + \mathcal{C}$, where $\tilde{\Lambda}$ has dimensions of mass, so that we can convert the $\ln\mu$ to a logarithm with a properly dimensionless argument. Next, since $\beta(a)$ is known only as a small-$a$ expansion, we naturally want the range of integration to be 0 to $a$. That gives us

$$\ln(\mu/\tilde{\Lambda}) = \int_0^a\frac{dx}{\beta(x)} + \mathcal{C}, \qquad (6.16)$$

but the integral is then divergent — so the constant $\mathcal{C}$ must also be divergent in a compensating fashion. A precise specification of $\mathcal{C}$ amounts to a definition of the $\tilde{\Lambda}$ parameter. (That will define $\tilde{\Lambda}$ within the specific RP we are using. Later on we will need to ask if $\tilde{\Lambda}$ depends on the RP choice.) One convenient choice of $\mathcal{C}$ is to write:

$$\ln(\mu/\tilde{\Lambda}) = \lim_{\delta\to 0}\left(\int_\delta^a\frac{dx}{\beta(x)} + \mathcal{C}(\delta)\right), \qquad (6.17)$$

with

$$\mathcal{C}(\delta) = \int_\delta^\infty\frac{dx}{bx^2(1 + cx)}. \qquad (6.18)$$

Note that $\mathcal{C}(\delta)$ involves only $b$ and $c$ and so is RS invariant.

**The tilde** in "$\tilde{\Lambda}$" is included to distinguish it from another widely-adopted definition of the $\Lambda$ parameter in the literature. The two definitions are related by an RS-invariant factor: $\ln(\Lambda/\tilde{\Lambda}) = (c/b)\ln|2c/b|$. See the discussion in section 6.4 below.

It is convenient to introduce the variable

$$\tau \equiv b \ln(\mu/\tilde{\Lambda}) \tag{6.19}$$

and to define

$$B(x) \equiv \frac{\beta(x)}{-bx^2} = 1 + cx + c_2 x^2 + \cdots . \tag{6.20}$$

Equation (6.17) then becomes, after splitting the $\mathcal{C}(\delta)$ integral into two pieces,

$$\tau = \lim_{\delta \to 0} \left( \int_\delta^a \frac{dx}{-x^2 B(x)} + \int_\delta^a \frac{dx}{x^2(1+cx)} + \int_a^\infty \frac{dx}{x^2(1+cx)} \right). \tag{6.21}$$

This equation gives $\tau$ as a function of $a$, which we shall name $K(a)$:

$$\tau = K(a) \equiv \int_a^\infty \frac{dx}{x^2(1+cx)} - \int_0^a \frac{dx}{x^2} \left( \frac{1}{B(x)} - \frac{1}{1+cx} \right). \tag{6.22}$$

Note that the second term is now a well-defined integral, convergent at the lower limit $x \to 0$. The first term gives the second-order approximation to $K(a)$ and is easily evaluated using partial fractions:

$$K^{(2)}(a) = \int_a^\infty \frac{dx}{x^2(1+cx)}$$
$$= \int_a^\infty dx \left( \frac{1}{x^2} - \frac{c}{x} + \frac{c^2}{1+cx} \right) \tag{6.23}$$
$$= \frac{1}{a} + c \ln \left| \frac{ca}{1+ca} \right| .$$

To summarize: the integrated $\beta$-function equation — which we will henceforth refer to as the "int-$\beta$ equation" — is given by

$$\tau = K(a), \tag{6.24}$$

with

$$K(a) = \frac{1}{a} + c \ln \left| \frac{ca}{1+ca} \right| - \Delta(a), \tag{6.25}$$

where

$$\Delta(a) \equiv \int_0^a \frac{dx}{x^2} \left( \frac{1}{B(x)} - \frac{1}{1+cx} \right). \tag{6.26}$$

The above form is quite satisfactory when $c$ is positive (which is the case for "real-world QCD," where $n_f \leq 6$). However, if $c$ is negative, which occurs in hypothetical theories with a large number of fermions, there is a slight problem with a pole at $x = -1/c$. In that case, we need to specify the Cauchy principal value for the $\mathcal{C}(\delta)$ integral in Eq. (6.18). The easiest way around the problem is to rewrite $K(a)$ in the form

$$K(a) = \frac{1}{a} + c \ln |ca| - \tilde{\Delta}(a), \tag{6.27}$$

where

$$\tilde{\Delta}(a) \equiv \int_0^a \frac{dx}{x^2} \left( \frac{1}{B(x)} - 1 + cx \right). \tag{6.28}$$

(For further discussion of this integral, see Appendix 9.A.) The equivalence between the two forms of $K(a)$ can easily be checked by verifying that

$$\tilde{\Delta}(a) - \Delta(a) = \int_0^a \frac{dx}{x^2} \left( -1 + cx + \frac{1}{1+cx} \right)$$

$$= \int_0^a dx \frac{c^2}{1+cx} \tag{6.29}$$

$$= c \ln |1 + ca|. \tag{6.30}$$

Thus, Eqs. (6.25) and (6.27) correspond to the *same* definition of $\tilde{\Lambda}$.

## 6.4. Inverting the int-$\beta$ Equation

Ideally, one would like to invert the int-$\beta$ equation to express the couplant $a$ as a function of $\tau$. Alas, that step cannot be carried out analytically. One might attempt to use an analytic approximation, such as an expansion in inverse powers of $\tau$, but that procedure has serious drawbacks. For one thing the expansion is ugly, involving logarithms of $\tau$. Also, it applies only when $\mu \gg \tilde{\Lambda}$ and thus precludes any attempt to go to low energies. Most seriously, the errors it introduces, even at high energies, are hard to estimate and control. The routine matter of inverting a transcendental equation should not be allowed to contribute significantly to the error in the final prediction. Instead, one should respect the $\tau = K(a)$ equation, solving it numerically whenever necessary, taking care that the numerical solution is accurate to several more decimal places than the minimum required for the final prediction to stay within its associated error estimate.

There is another important reason not to use a series in powers of $1/\ln(\mu/\tilde{\Lambda})$: truncating such a series introduces a spurious dependence on exactly how one chooses to define $\tilde{\Lambda}$. The definition of $\tilde{\Lambda}$ made earlier corresponds to the particular choice for the constant of integration $\mathcal{C}(\delta)$ in Eq. (6.18). That choice is quite elegant and convenient, but it is not unique and is just a convention. One could well have chosen some other $\mathcal{C}(\delta)$, differing by a finite, RS-invariant number. Indeed, the historical definition of $\Lambda$ (without the tilde) is an example of a different choice. The existence of different conventions for how to define the $\Lambda$ or $\tilde{\Lambda}$ parameters creates no problem, since one can convert *exactly* between the different definitions. Using truncated series in $1/\ln(\mu/\Lambda)$ or $1/\ln(\mu/\tilde{\Lambda})$ would spoil that exactness and introduce a new source of ambiguity. That ambiguity is wholly avoidable, and it is sensible to avoid it.

**The original definition** of the $\Lambda$ parameter, by Buras *et al.*, was such that the series expansion of the couplant $a$ as a series in $1/\ln(\mu^2/\Lambda^2)$ had the form (in the original notation)

$$g^2 = \frac{1}{\beta_0 \ln(\mu^2/\Lambda^2)} - \frac{[\beta_1 \ln(\ln(\mu^2/\Lambda^2)) + C]}{\beta_0^3 \ln^2(\mu^2/\Lambda^2)} + \cdots ,$$

with the constant $C$ chosen to be zero. Converting to our notation $a = g^2/(4\pi^2)$, $b = 8\pi^2\beta_0$, $c = 4\pi^2\beta_1/\beta_0$, one can check that this corresponds to an approximate inversion of

$$\frac{1}{a} + c\ln\left|\frac{ba}{2}\right| + O(a) = b\ln(\mu/\Lambda),$$

instead of our form

$$K(a) = \frac{1}{a} + c\ln|ca| + O(a) = b\ln(\mu/\tilde{\Lambda}),$$

which follows from Eqs. (6.25) or (6.27), because $\Delta(a)$ or $\tilde{\Delta}(a)$ are $O(a)$. Comparing the above equations shows that $\ln(\Lambda/\tilde{\Lambda}) = (c/b)\ln|2c/b|$, as stated earlier.

## 6.5. RP Dependence of $\tilde{\Lambda}$: The CG Relation

Having defined $\tilde{\Lambda}$ we should now ask; is it RP dependent? The answer is "yes," but in a simple and definite way, described by the Celmaster–Gonsalves (CG) relation.

**Theorem (Celmaster and Gonsalves).** *If two prescriptions (two schemes with the same value of $\mu$) are related by*

$$a' = a(1 + v_1 a + \cdots),  \tag{6.31}$$

*then*

$$\ln(\tilde{\Lambda}'/\tilde{\Lambda}) = v_1/b.  \tag{6.32}$$

*This result is* exact *and does not involve the $v_2, v_3, \ldots$ coefficients.*

**Proof.** We start from Eq. (6.17) and the corresponding equation in a different, primed, RS. Subtracting those two equations, with $\mu' = \mu$, yields

$$\ln(\tilde{\Lambda}'/\tilde{\Lambda}) = \lim_{\delta \to 0} \left( \int_\delta^a \frac{dx}{\beta(x)} - \int_\delta^{a'} \frac{dx'}{\beta'(x')} \right).  \tag{6.33}$$

Evaluating for small $a$ leads to

$$\ln(\tilde{\Lambda}'/\tilde{\Lambda}) = \frac{1}{b}\left(\frac{1}{a} - \frac{1}{a'}\right) + O(a) = v_1/b + O(a).  \tag{6.34}$$

Now, since the left-hand side is independent of $\mu$, we can choose to evaluate the right-hand side at any value of $\mu$. Choosing $\mu \to \infty$ (if $b$ is positive), or $\mu \to 0$ (if $b$ is negative), makes $a \to 0$ and so completes the proof.

This proof, though it is sound, can seem too slick. The following alternative proof (due to Osborn) is instructive. First, one splits each of the integrals in Eq. (6.33) into two pieces:

$$\ln(\tilde{\Lambda}'/\tilde{\Lambda}) = \lim_{\delta \to 0} \left( \int_{\delta}^{\epsilon} \frac{dx}{\beta(x)} + \int_{\epsilon}^{a} \frac{dx}{\beta(x)} - \int_{\delta}^{\epsilon'} \frac{dx'}{\beta'(x')} - \int_{\epsilon'}^{a'} \frac{dx'}{\beta'(x')} \right).$$
(6.35)

Then one chooses $\epsilon'$ to be related to $\epsilon$ by

$$\epsilon' = \epsilon(1 + v_1\epsilon + \cdots),$$
(6.36)

with exactly the same coefficients as in Eq. (6.31). The second and fourth terms in (6.35) can then be shown to cancel exactly by making the change of variables

$$x' = x(1 + v_1x + \cdots),$$
(6.37)

and using the relation of $\beta'$ to $\beta$ of Eq. (6.8). One can now take the limit $\epsilon \to 0$ (choosing, say, $\epsilon = 2\delta$ or $\epsilon = \sqrt{\delta}$) and use the fact that $\beta(x) \sim -bx^2$ as $x \to 0$ to obtain the result (6.32). This proof makes plain that the result is non-zero only because of the $1/x^2$ singularity of the $1/\beta$ integrands and that the behaviour of $\beta(x)$ away from the infinitesimal neighbourhood of $x = 0$ plays no role.

The importance of the Celmaster–Gonsalves relation is its exactness. It means that, although the $\tilde{\Lambda}$ parameter is prescription dependent, the $\tilde{\Lambda}$'s of different prescriptions can be related exactly by a straightforward Feynman-diagram calculation of the single coefficient $v_1$ in the relation between the two prescriptions. Hence, the $\tilde{\Lambda}$ parameter of any convenient "reference RP" can be adopted, without prejudice, as the one free parameter of the theory, taking over the role of the "bare coupling constant" in the theory's original Lagrangian. This allows us to do the proper "book-keeping" when comparing theory and experiment for many different physical quantities. If two people choose to use different RP's as the "reference RP," that is not a problem: one can convert exactly between the $\tilde{\Lambda}$'s of the two conventions — and that is true in a practical sense, not just in principle.

**By contrast,** reporting experimental results fitted to QCD predictions in terms of an extracted numerical value for "$\alpha_s^{\overline{\text{MS}}}(M_Z)$" creates a quagmire of unnecessary ambiguities. The $\overline{\text{MS}}$ scheme, while convenient for diagrammatic calculations, is an arbitrary choice and leaves open the question of how to fix $\mu$. The conversion from $a_{\overline{\text{MS}}}(M_Z)$ to the couplant of another RP can only be made approximately, with an uncontrolled error — for instance, does one express the new $a$ as a truncated series in $a_{\overline{\text{MS}}}$, or the other way around? Even to convert the scale from $M_Z$ (the $Z$-boson mass) to some other scale is ambiguous: It is not clear how many terms of the $\overline{\text{MS}}$ $\beta$ function should be used, especially when the fit has been made to various quantities, some of which are known to higher orders than others. For systematic comparisons of theory and experiment it would be sensible to use, say, $\tilde{\Lambda}_{\overline{\text{MS}}}$ as the free parameter of QCD. If one ever wanted to convert to the $\tilde{\Lambda}$ parameter of a different reference RP then that conversion could be done *exactly*, with no ambiguity.

Of course, the value of $\tilde{\Lambda}_{\overline{\text{MS}}}$ extracted from even very precise experiments tends to have a large error estimate, so one might diplomatically adopt the convention to quote, say, the value of $\ln(1\,\text{GeV}/\tilde{\Lambda}_{\overline{\text{MS}}})$.

**Exercise 6.1.** Consider a "delicate" theory in which $b = 0$, so that the $\beta$ function starts at order $a^3$:

$$\mu\frac{da}{d\mu} = \beta(a) = ha^3(1 + g_1 a + g_2 a^2 + \cdots).$$

(Assume $h$ is non-zero.)

(i) Repeat the 't Hooft argument in Sec. 6.2 to show that $h$ and $g \equiv g_2 - g_1^2$ are invariants.
(ii) Show that a suitable constant of integration for the int-$\beta$ equation (6.17) (with $\lim_{\delta\to 0}$ understood) is

$$\mathcal{C}(\delta) = -\frac{1}{h}\left(\frac{1}{2\delta^2} - \frac{g_1}{\delta} + g\ln\delta\right).$$

Note that this $\mathcal{C}(\delta)$ is now RS dependent (unlike the usual case), though in a simple way: $\partial\mathcal{C}/\partial g_1 = 1/(h\delta)$.
(iii) Show that the equivalent to the CG relation is

$$h\ln(\tilde{\Lambda}'/\tilde{\Lambda}) = -v_2 + g_1 v_1 + \frac{1}{2}v_1^2,$$

so that the $\tilde{\Lambda}$'s of different RP's are related exactly by a 2-loop calculation.

(iv)  Revisit Exercise 3.2 and show that for a "delicate" theory the
't Hooft analysis implies that the bare coupling constant $a_B$
is of order $\sqrt{\epsilon}$.

(See also Exercise 7.7 at the end of the next chapter.)

# Chapter 7

# Parametrization of RS Dependence and the $\rho_n$ Invariants

## 7.1. $\beta$ Function Coefficients as Scheme Labels

To make progress we now need to identify the "extraneous variables" associated with RS dependence: That is, we need a parametrization of all possible choices of RS. At first sight, the task might appear hopeless: There are all sorts of ways to define an RS, and they usually relate to very technical details of Feynman-diagram calculations, and involve lots of arbitrary choices and conventions. Looked at in the right way, however, the problem of how to parametrize RS's has a quite simple solution.

First, recall the int-$\beta$ equation, (6.17):

$$\ln(\mu/\tilde{\Lambda}) = \lim_{\delta \to 0} \left( \int_{\delta}^{a} \frac{dx}{\beta(x)} + \mathcal{C}(\delta) \right) \equiv \text{ `` } \int_{[0]}^{a} \frac{dx}{\beta(x)} \text{ '' }. \qquad (7.1)$$

Since the integration constant $\mathcal{C}(\delta)$ is RS invariant, it is clear that $a$ can depend on RS only through $\tau \equiv b \ln(\mu/\tilde{\Lambda})$ and the scheme-dependent coefficients $c_2, c_3, \ldots$ of the $\beta$ function, because these are the only RS-dependent quantities (besides $a$ itself) involved in the equation. The next step is to observe that a physical quantity $\mathcal{R}$ cannot possibly depend on any *other* RS parameters, because the cancellation expressed by the symbolic equation (6.2) could not occur

---

if the $\mathcal{R}$ coefficients depended on some RS variable that $a$ did not depend on. Thus, as far as physical quantities are concerned, these variables must provide a complete RS parametrization:

$$\text{RS} = \{\tau, c_2, c_3, \ldots\}. \tag{7.2}$$

Note the important point that $\mu$ and $\tilde{\Lambda}$ do not enter separately, but only through $\tau \equiv b \ln(\mu/\tilde{\Lambda})$.

The importance of this simple argument takes some time to appreciate, but it is crucial for all that follows. The reader is urged to re-read the preceding paragraph carefully and critically.

> **Note that** the claim that the parametrization by $\tau, c_2, c_3, \ldots$ is *complete* is qualified by saying *as far as physical quantities are concerned*: Non-physical quantities, such as Green's functions, can depend on other aspects of the renormalization procedure. That is because Green's functions depend on the wavefunction-renormalization constant, which cancels out in physical quantities. Equivalently, we can say that a Green's function $G$ has an "anomalous dimension," $\frac{\mu}{G}\frac{dG}{d\mu}$, that will depend on the renormalization procedure in ways other than through $\tau, c_2, c_3, \ldots$. This point is relevant to the problem of factorization-scheme dependence, discussed in Chapter 12.

> **Because** $\mu$ and $\tilde{\Lambda}$ do not enter separately, but only through $\tau \equiv b \ln(\mu/\tilde{\Lambda})$, it is neither necessary nor sensible to fix, separately, the RP and the value of $\mu$ in order to fix the RS. For instance, two RP's that have the same $c_2, c_3, \ldots$ and differ only in their value of $\tilde{\Lambda}$, could be made entirely equivalent simply by redefining the $\mu$ parameter of one of them. (MS and $\overline{\text{MS}}$ are two such RP's.) As shown in the next chapter, "optimization" determines the optimal $\tau$ for a given quantity at a given order, but it does not determine an "optimal $\mu$;" nor is one needed. (More informally, one might speak of an "optimal $\mu$," if it is understood as relative to the $\tilde{\Lambda}$ parameter of some specified reference RP.)

> **In gauge theories** some RP's, particularly momentum-subtraction schemes, are "gauge dependent." One might worry that the parametrization by $\tau, c_2, c_3, \ldots$ is then somehow incomplete, but that it not so. In such cases, the RP is not fully specified until the gauge choice is made. Two such prescriptions, differing only in gauge choice, are not a single gauge-dependent RP, but are just

two different RP's — and they are labelled by different values of $c_2, c_3, \ldots$.

The parametrization of RS dependence by the set of variables $\tau, c_2, c_3, \ldots$ is not the only possible parametrization (see Exercise 7.2), but it does provide a convenient "Cartesian coordinate system." The partial derivative with respect to any one of these variables is to be taken holding the other variables of the set constant. That is, $\partial/\partial\tau$ is taken with all the $c_i$'s held constant, which corresponds to varying the renormalization scale $\mu$ while holding the RP constant. Similarly, $\partial/\partial c_j$ is taken with $\tau$ and the other $c_i$ ($i \neq j$) held constant.

The symbolic RG-invariance equation (6.2) can now be written out explicitly as the following set of equations:

$$\frac{\partial \mathcal{R}}{\partial \tau} = \left( \left.\frac{\partial}{\partial \tau}\right|_a + \frac{\beta(a)}{b}\frac{\partial}{\partial a} \right) \mathcal{R} = 0,$$

$$\frac{\partial \mathcal{R}}{\partial c_j} = \left( \left.\frac{\partial}{\partial c_j}\right|_a + \beta_j(a)\frac{\partial}{\partial a} \right) \mathcal{R} = 0, \quad j = 2, 3, \ldots,$$
(7.3)

involving some new functions $\beta_j(a)$ defined by

$$\beta_j(a) \equiv \frac{\partial a}{\partial c_j}.$$
(7.4)

These $\beta_j(a)$ functions are fixed in terms of the $\beta(a)$ function, as we show in the next section.

## 7.2. The $\beta_j(a)$ and $B_j(a)$ Functions

Consider the int-$\beta$ equation in the form (7.1). The left-hand side is $\tau/b$ and depends only on $\tau$, not on the $c_j$'s. The constant $\mathcal{C}(\delta)$ is also independent of the $c_j$'s. Therefore, applying the partial derivative $\frac{\partial}{\partial c_j}$ to the equation leads to

$$0 = \frac{\beta_j(a)}{\beta(a)} + \int_0^a dx \left(\frac{-1}{\beta(x)^2}\right)\left(-bx^{j+2}\right),$$
(7.5)

where the first term arises from the $c_j$ dependence of the limit of integration $a$, and the second term arises from the explicit $c_j$

dependence of the integrand. Rearranging this equation gives us

$$\beta_j(a) = -b\beta(a) \int_0^a dx \, \frac{x^{j+2}}{\beta(x)^2}, \tag{7.6}$$

for $j = 2, 3, \ldots$.

The $\beta_j$ functions begin at order $a^{j+1}$ (see Appendix 7.A for their series expansions). It is convenient to define $B_j(a)$ functions whose series expansions begin $1 + O(a)$:

$$B_j(a) \equiv \frac{(j-1)}{a^{j+1}} \beta_j(a). \tag{7.7}$$

For $j = 1$, it is natural to define

$$B_1(a) \equiv B(a) \equiv \frac{\beta(a)}{-ba^2} = 1 + ca + c_2 a^2 + \cdots = \sum_{i=0}^{\infty} c_i a^i, \tag{7.8}$$

with the convention that $c_0 \equiv 1$ and $c_1 \equiv c$. Equation (7.6) can then be rewritten as

$$B_j(a) = \frac{(j-1)}{a^{j-1}} B(a) I_j(a), \tag{7.9}$$

where

$$I_j(a) \equiv \int_0^a dx \, \frac{x^{j-2}}{B(x)^2}. \tag{7.10}$$

(Note that this formula for $B_j(a)$ even holds for $j = 1$ if the right-hand side is interpreted as the limit $j \to 1$ from above. See also Exercise 7.3.)

The $B_j(a)$ functions have the power-series expansions

$$B_j(a) \equiv \sum_{i=0}^{\infty} W_i^j a^i, \tag{7.11}$$

with $W_0^j \equiv 1$. The other $W_i^j$ coefficients are fixed in terms of the $c_i$'s. A convenient formula expressing that fact can be obtained as follows. Differentiating Eq. (7.6) with respect to $a$ (holding the $c_j$ coefficients constant) yields the differential equation

$$\beta_j'(a)\beta(a) - \beta_j(a)\beta'(a) = -ba^{j+2}. \tag{7.12}$$

(This equation may also be obtained by considering the commutation of the second derivatives, $\partial^2 a/\partial\tau\partial c_j = \partial^2 a/\partial c_j\partial\tau$; see Exercise 7.1.) In terms of the $B$ and $B_j$ functions this becomes

$$(j-1)B_j B + a(B_j' B - B_j B') = (j-1). \tag{7.13}$$

Equating powers of $a$ leads to the relation

$$\sum_{m=0}^{i}(i+j-1-2m)c_m W_{i-m}^j = (j-1)\delta_{i0} \tag{7.14}$$

for $i = 0, 1, 2, \ldots$ and $j = 1, 2, \ldots$. In the special case $j = 1$, one has $W_i^1 \equiv c_i$ and the above equation reduces to

$$\sum_{m=0}^{i}(i-2m)c_m c_{i-m} = 0, \tag{7.15}$$

which is true identically, since the left-hand side is

$$\sum_{m=0}^{i}(i-m)c_m c_{i-m} - \sum_{m=0}^{i} m c_m c_{i-m} \tag{7.16}$$

and the first sum, by changing the summation variable from $m$ to $n = i - m$, is seen to cancel the second.

## 7.3. The $\rho_n$ Invariants

The RG equations (7.3) determine how the coefficients $r_i$ of $\mathcal{R}$ must depend on the RS variables $\{\tau, c_j\}$. To show explicitly how this works we specialize to the $\mathrm{P} = 1$ case, where

$$\mathcal{R} = a(1 + r_1 a + r_2 a^2 + \cdots) \tag{7.17}$$

and write out the lowest-order terms to obtain

$$\left(a^2\frac{\partial r_1}{\partial\tau} + a^3\frac{\partial r_2}{\partial\tau} + \cdots\right) - a^2(1+ca+\cdots)(1+2r_1 a+\cdots) = 0, \tag{7.18}$$

$$\left(a^2\frac{\partial r_1}{\partial c_2} + a^3\frac{\partial r_2}{\partial c_2} + \cdots\right) + a^3(1+W_1^2 a+\cdots)(1+2r_1 a+\cdots) = 0. \tag{7.19}$$

and so on. (In fact, the coefficient $W_1^2$ is zero.) Equating powers of $a$, one sees that $r_1$ depends on $\tau$ only, while $r_2$ depends on $\tau$ and $c_2$ only, etc., with

$$\frac{\partial r_1}{\partial \tau} = 1, \tag{7.20}$$

$$\frac{\partial r_2}{\partial \tau} = 2r_1 + c, \qquad \frac{\partial r_2}{\partial c_2} = -1, \tag{7.21}$$

etc. Upon integration one will obtain $r_i$ as a function of $\tau, c_2, \ldots, c_i$ plus a constant of integration that is RS invariant. Thus, certain combinations of series coefficients and RS parameters are RS invariant. From Eqs. (7.20) and (7.21), one sees that

$$\boldsymbol{\rho}_1(Q) \equiv \tau - r_1 \tag{7.22}$$

and

$$\rho_2 \equiv c_2 + r_2 - cr_1 - r_1^2 \tag{7.23}$$

are invariants.

The first invariant, $\boldsymbol{\rho}_1(Q)$, is unique in being dependent on the physical energy scale, $Q$. (Recall the discussion in Sec. 6.1, and Chapters 2 and 3.) A QFT calculation of the coefficient $r_1$, in some arbitrary RS, yields a result of the form

$$r_1 = b \ln(\mu/Q) + r_{1,o}, \tag{7.24}$$

whose $\mu$ dependence indeed conforms with Eq. (7.20). For dimensional reasons, the $\mu$ and $Q$ dependences are tied together in $r_1$. (The QFT calculation of $r_1$ does not "know" what boundary condition will later be applied to the $\beta$-function equation, so the parameter $\tilde{\Lambda}$ cannot explicitly appear.) Similarly, the higher coefficients $r_2, \ldots$ depend on $\ln(\mu/Q)$, but not on $\mu$ or $Q$ separately. Hence, for the invariants $\rho_2, \rho_3, \ldots$ the cancellation of $\mu$ dependence also implies the cancellation of $Q$ dependence. However, $\boldsymbol{\rho}_1(Q)$ is different because its definition explicitly involves $\tau$:

$$\boldsymbol{\rho}_1(Q) = b \ln(\mu/\tilde{\Lambda}) - (b \ln(\mu/Q) + r_{1,o})$$
$$= b \ln(Q/\tilde{\Lambda}) - r_{1,o}, \tag{7.25}$$

where $\tilde{\Lambda}$ is the $\tilde{\Lambda}$-parameter of the RP in which the QFT calculation of $r_{1,o}$ was done. Note that $\rho_1(Q)$ is *both* $\mu$ independent *and* RP independent: If $a$ is changed to $a(1+v_1a+\cdots)$ then $r_{1,o}$ changes, but so does $b\ln\tilde{\Lambda}$ by a compensating amount, thanks to the Celmaster–Gonsalves relation. Indeed, we may write

$$\rho_1(Q) = b\,\ln(Q/\tilde{\Lambda}_{\mathcal{R}}), \tag{7.26}$$

where $\tilde{\Lambda}_{\mathcal{R}}$ is a characteristic scale specific to the particular physical quantity $\mathcal{R}$. Knowing $r_{1,o}$, we can relate it exactly to the $\tilde{\Lambda}$ of the original prescription.

Some convention must be adopted to uniquely define the higher invariants $\rho_j$ (for $j \geq 2$) because, of course, any sum of invariants is also an invariant. For example, one might quite naturally add some multiple of $c^2$ to Eq. (7.23).

**Indeed,** a different definition was employed in the author's early papers, with $\rho_2^{\text{old}} = \rho_2 - \frac{1}{4}c^2$ and $\rho_3^{\text{old}} = \frac{1}{2}\rho_3$. In later papers, the definition described below was used and distinguished by a tilde ("$\tilde{\rho}_j$"). Here we shall dispense with the tildes.

To define the $\rho_j$'s we proceed as follows: For any given physical quantity $\mathcal{R}$, one can always define an RS (known either as the "fastest apparent convergence" (FAC) or "effective charge" (EC) scheme) such that all the series coefficients $r_i$ vanish in that scheme, so that $\mathcal{R} = a_{\text{EC}}(1+0+0+\cdots)$. Since the $\beta$ functions of any two RS's are related by Eq. (6.8), we must have

$$\beta_{\text{EC}}(\mathcal{R}) = \frac{\partial\mathcal{R}}{\partial a}\beta(a). \tag{7.27}$$

The $\rho_n$ invariants are defined to coincide with the coefficients of the EC-scheme $\beta$ function:

$$\beta_{\text{EC}}(\mathcal{R}) \equiv -b\mathcal{R}^2\sum_{n=0}^{\infty}\rho_n\mathcal{R}^n = -ba^2\frac{\partial\mathcal{R}}{\partial a}B(a), \tag{7.28}$$

where $\rho_0 \equiv 1$ and $\rho_1 \equiv c$ (not to be confused with the independent invariant $\rho_1(Q) \equiv \tau - r_1$, which will always be written with its $Q$

argument). Rearranging this equation as

$$B(a) = \sum_{n=0}^{\infty} \rho_n a^n \left(\frac{\mathcal{R}}{a}\right)^{n+2} \frac{1}{\frac{\partial \mathcal{R}}{\partial a}} \qquad (7.29)$$

and then equating powers of $a$ yields

$$c_j = \sum_{i=0}^{j} \rho_i \mathbb{C}_{j-i}\left[\left(\frac{\mathcal{R}}{a}\right)^{i+2} \frac{1}{\frac{\partial \mathcal{R}}{\partial a}}\right], \qquad (7.30)$$

where $\mathbb{C}_n[F(a)]$ means "the coefficient of $a^n$ in the series expansion of $F(a)$."

The first few invariants, for $P = 1$, are as follows:

$$\rho_1 = c, \quad \text{and} \quad \rho_1(Q) = \tau - r_1,$$
$$\rho_2 = c_2 - cr_1 + r_2 - r_1^2,$$
$$\rho_3 = c_3 - 2c_2 r_1 + cr_1^2 + 4r_1^3 - 6r_1 r_2 + 2r_3, \qquad (7.31)$$
$$\rho_4 = c_4 - 3c_3 r_1 + c_2(4r_1^2 - r_2) + c(-2r_1 r_2 + r_3)$$
$$- 14r_1^4 + 28r_1^2 r_2 - 5r_2^2 - 12r_1 r_3 + 3r_4.$$

The generalization to any $P$ is as follows:

$$\rho_1 = c, \quad \text{and} \quad \rho_1(Q) = \tau - \frac{r_1}{P},$$

$$\rho_2 = c_2 - c\frac{r_1}{P} + \frac{r_2}{P} - \frac{(P+1)}{2P^2}r_1^2,$$

$$\rho_3 = c_3 - 2c_2\frac{r_1}{P} + c\frac{r_1^2}{P^2} + \frac{2(P+1)(P+2)}{3P^3}r_1^3 - \frac{2(P+2)}{P^2}r_1 r_2 + \frac{2r_3}{P},$$
$$\qquad (7.32)$$

$$\rho_4 = c_4 - 3c_3\frac{r_1}{P} + c_2\left(\frac{(P+7)}{2P^2}r_1^2 - \frac{r_2}{P}\right)$$

$$+ c\left(\frac{(P-1)(2P+5)}{6P^3}r_1^3 - \frac{(P+1)}{P^2}r_1 r_2 + \frac{r_3}{P}\right)$$

$$-\frac{(\mathrm{P}+1)(2\mathrm{P}+5)(3\mathrm{P}+5)}{8\mathrm{P}^4}r_1^4 + \frac{(2\mathrm{P}+5)(3\mathrm{P}+5)}{2\mathrm{P}^3}r_1^2 r_2$$

$$-\frac{(3\mathrm{P}+7)}{2\mathrm{P}^2}r_2^2 - \frac{3(\mathrm{P}+3)}{\mathrm{P}^2}r_1 r_3 + \frac{3r_4}{\mathrm{P}}.$$

Equation (7.30) generalizes to

$$c_j = \sum_{i=0}^{j} \rho_i \mathbb{C}_{j-i} \left[\left(\frac{\mathcal{R}}{a^{\mathrm{P}}}\right)^{(i+\mathrm{P}+1)/\mathrm{P}}\frac{1}{\mathcal{S}}\right], \tag{7.33}$$

where $\mathcal{S} \equiv \frac{1}{\mathrm{P}a^{\mathrm{P}-1}}\frac{\partial \mathcal{R}}{\partial a}$. An inverse formula giving the $\rho_n$'s in terms of the $c_j$ and $r_j$ coefficients is

$$\rho_n = \mathbb{C}_n \left[B(a)\left(\frac{\mathcal{R}}{a^{\mathrm{P}}}\right)^{-(n+2\mathrm{P}+1)/\mathrm{P}}\mathcal{S}^2\right]. \tag{7.34}$$

## Appendix 7.A: Expansions of the $\beta_j(a)$ Functions

The opening terms of the series expansions of the first few $\beta_j(a)$ functions are given below. Note that the $B_j(a)$ functions are given by the series in parentheses, whose coefficients are the $W_i^j$ of Eq. (7.11):

$$\beta_2(a) = a^3\left(1 + 0 + \frac{c_2}{3}a^2 + \frac{3c_3 - cc_2}{6}a^3\right.$$
$$\left. + \frac{18c_4 - 9cc_3 - 2c_2^2 + 3c^2 c_2}{30}a^4 + O(a^5)\right),$$

$$\beta_3(a) = \frac{1}{2}a^4\left(1 - \frac{c}{3}a + \frac{c^2}{6}a^2 + \frac{6c_3 + 2cc_2 - 3c^3}{30}a^3 + O(a^4)\right),$$

$$\beta_4(a) = \frac{1}{3}a^5\left(1 - \frac{c}{2}a + \frac{3c^2 - 2c_2}{10}a^2 + O(a^3)\right),$$

$$\beta_5(a) = \frac{1}{4}a^6\left(1 - \frac{3c}{5}a + O(a^2)\right),$$

$$\beta_6(a) = \frac{1}{5}a^7\left(1 + O(a)\right).$$

These expansions of $\beta_j(a)$ are useful for formal purposes. However, the approximate forms needed later are not truncated series, but correspond to the definition, Eq. (7.6), with $\beta(a)$ replaced by its truncated series.

**Exercise 7.1.** Show that requiring commutation of the second derivatives, $\partial^2 a/\partial\tau\partial c_j = \partial^2 a/\partial c_j\partial\tau$, leads to

$$\beta_j'(a)\beta(a) - \beta_j(a)\beta'(a) = -ba^{j+2},$$

where the prime indicates differentiation with respect to $a$, regarding the coefficients $c_j$ as fixed. (Note that the coefficients of $\beta_j(a)$ cannot depend on $\mu$ by dimensional analysis.) Verify that the solution of this differential equation is Eq. (7.6).

**Exercise 7.2.** Consider a more general way of labelling RS's than the $\tau, c_2, c_3, \ldots$ "coordinate system." Make the following minimal set of assumptions about these RS labels $u_1, u_2, u_3, \ldots$: (i) the dependence of $a$ on $u_j$ starts at order $a^{j+1}$, and (ii) these parameters are mutually independent. Also assume that $u_1 = \tau$, which loses no real generality since we know that $\partial a/\partial\tau = \beta(a)/b$ starts at order $a^2$, and that the renormalization scale $\mu$ must be involved in the RS labelling. Assumption (i) implies that

$$\frac{\partial a}{\partial u_j} = N_j a^{j+1}\left(1 + \tilde{W}_1^j a + \tilde{W}_2^j a^2 + \cdots\right),$$

with some normalization $N_j$ and coefficients $\tilde{W}_i^j$. From assumption (ii) we must have

$$\frac{\partial^2 a}{\partial\tau\partial u_j} = \frac{\partial^2 a}{\partial u_j\partial\tau}.$$

Show that this leads to the conclusion that

$$\frac{\partial c_{j-i}}{\partial u_j} = 0,$$

$$\frac{\partial c_j}{\partial u_j} = N_j(j-1),$$

and

$$\frac{\partial c_{j+i}}{\partial u_j} = N_j\sum_{r=0}^{i}(i+j-1-2r)\,c_r\tilde{W}_{i-r}^j,$$

for $j = 2, 3, \ldots$ and $i = 1, 2, \ldots$, with $c_1 \equiv c$ and $c_0 \equiv \tilde{W}_0^j \equiv 1$.

Note the important conclusion that $c_j$ is linearly dependent on $u_j$. Check that if one chooses $u_j = c_j$ then the $\tilde{W}_i^j$ coefficients reduce to the $W_i^j$ coefficients of the $B_j(a)$ functions. See Eqs. (7.11) and (7.14).

**Exercise 7.3.** Show that the RG equations for the $c_j$'s, written in terms of the $B_j(a)$ functions

$$\left( \left. \frac{\partial}{\partial c_j} \right|_a + \frac{a^{j+1} B_j(a)}{(j-1)} \frac{\partial}{\partial a} \right) \mathcal{R} = 0,$$

can be taken to apply even for $j = 1$ if we consider "$c_1$" to be the invariant $c$ plus an infinitesimal part proportional to $\tau$, in the sense that

$$\text{"}c_1\text{"} = \lim_{j \to 1} (c - (j-1)\tau)$$

so that

$$\frac{\partial}{\partial c_1} \to -\frac{1}{(j-1)} \frac{\partial}{\partial \tau}.$$

(Note that, as discussed in Sec. 7.2, the formula for $B_j$ naturally leads to the identification of $B_1(a)$ with $B(a)$.)

**Exercise 7.4.** Find the first three coefficients in the conversion between the couplants of two RS's, $a$ and $\tilde{a}$:

$$\tilde{a} = a(1 + V_1 a + V_2 a^2 + V_3 a^3 + \cdots).$$

The two RS's are labelled by $\tau, c_2, c_3, \ldots$ and $\tilde{\tau}, \tilde{c}_2, \tilde{c}_3, \ldots$, respectively, where $\tau = b \ln(M/\tilde{\Lambda})$ and $\tilde{\tau} = b \ln(\tilde{M}/\tilde{\Lambda})$, and $M$ and $\tilde{M}$ are the renormalization-scale choices in the two schemes. (Without loss of generality, we can assume that the two RP's are defined so that their $\tilde{\Lambda}$'s are the same.)

**Solution:** The result will be needed in Chapter 12. $V_1$ is easily found from considering the $\tau$ and $\tilde{\tau}$ derivatives. The other coefficients can best be found from the relation (see Eq. (6.8)) between the two $\beta$ functions:

$$\tilde{\beta}(\tilde{a}) = \frac{d\tilde{a}}{da} \beta(a).$$

Indeed, the calculation is algebraically the same as that for finding the $\rho$ invariants from Eq. (7.27), with the replacements $\mathcal{R} \to \tilde{a}$, $\rho_j \to \tilde{c}_j$, and $r_i \to V_i$. Thus, we may use Eq. (7.30) with those replacements, and then rearrange to solve for $V_i$. The first three are

$$V_1 = \tau - \tilde{\tau},$$
$$V_2 = (\tau - \tilde{\tau})^2 + c(\tau - \tilde{\tau}) - (c_2 - \tilde{c}_2),$$
$$V_3 = (\tau - \tilde{\tau})^3 + \frac{5}{2}c(\tau - \tilde{\tau})^2 + (-2c_2 + 3\tilde{c}_2)(\tau - \tilde{\tau}) - \frac{1}{2}(c_3 - \tilde{c}_3).$$

Note that $V_i$ is independent of all $c_j$'s with $j > i$. Note also that the $\tau$ and $\tilde{\tau}$ dependence is only through the difference $\tau - \tilde{\tau}$, so that the $V_i$'s depend on $M$ and $\tilde{M}$ only through the dimensionless ratio $M/\tilde{M}$, as required dimensionally. However, note that the dependence on the $c_j$, $\tilde{c}_j$ is *not* always through the difference $c_j - \tilde{c}_j$. This fact, first seen in the $c_2$, $\tilde{c}_2$ dependence of $V_3$, might seem oddly asymmetrical, but is in fact necessary for overall $a \leftrightarrow \tilde{a}$ symmetry. Reversion of a power series gives the inverse relation:

$$a = \tilde{a}(1 + \tilde{V}_1 \tilde{a} + \tilde{V}_2 \tilde{a}^2 + \tilde{V}_3 \tilde{a}^3 + \cdots),$$

with

$$\tilde{V}_1 = -V_1$$
$$\tilde{V}_2 = -V_2 + 2V_1^2,$$
$$\tilde{V}_3 = -V_3 + 5V_2 V_1 - 5V_1^3.$$

One can check that, after substituting the above results for $V_1, V_2, V_3$, one indeed finds that $\tilde{V}_1, \tilde{V}_2, \tilde{V}_3$ are given by those same expressions with all tilde and plain variables exchanged.

**Exercise 7.5.** Consider the successive logarithmic derivatives of a physical quantity $\mathcal{R} = a^P(1 + r_1 a + \cdots)$ with respect to the physical energy scale $Q$:

$$\mathcal{R}_{[n+1]} \equiv Q\frac{d\mathcal{R}_{[n]}}{dQ},$$

for $n = 1, 2, 3, \ldots$, where $\mathcal{R}_{[1]} \equiv \mathcal{R}$. These must all be physical quantities themselves.

(i) In a fixed RS, the $Q$ dependence comes only from the $r_i$ coefficients' dependence on $Q/\mu$. Use the $\mu$ RG equation to show that

$$\mathcal{R}_{[2]} = \beta(a)\frac{\partial \mathcal{R}}{\partial a}.$$

(ii) Apply the result again to get $\mathcal{R}_{[3]}$ and hence show that the quantity

$$\gamma \equiv \frac{\mathcal{R}_{[3]}}{\mathcal{R}_{[2]}} = 1 + Q \frac{d^2\mathcal{R}}{dQ^2} \bigg/ \frac{d\mathcal{R}}{dQ}$$

is given by

$$\gamma = \frac{d\beta}{da} + \beta(a)\frac{\partial^2\mathcal{R}}{\partial a^2} \bigg/ \frac{\partial\mathcal{R}}{\partial a}.$$

($\gamma$ is an "effective exponent" in the sense that $\mathcal{R}$ is locally described, in the near neighbourhood of some particular energy $Q$, by a power law $\mathcal{R} \approx \text{const.} + CQ^\gamma$. It is particularly interesting in the infrared limit, as we discuss in Chapter 11.)

**Exercise 7.6.** (i) Continuing from the previous exercise, find the first few perturbative coefficients of $\mathcal{R}_{[2]}$ and hence find the invariants associated with it. Show that these are simply combinations of the usual $\rho_1(Q)$ and $\rho_n$ invariants for the original $\mathcal{R}$. (Note that $\mathcal{R}_{[2]}$ has a different leading-order power from $\mathcal{R}$, namely P + 1 instead of P.)

(ii) The result is rather simpler for $\mathcal{R}_{[2]}/\mathcal{R}$, which is the scale dimension of $\mathcal{R}$:

$$\mathcal{D}_{(\mathcal{R})} \equiv \frac{Q}{\mathcal{R}}\frac{d\mathcal{R}}{dQ}.$$

Show that the invariants for $\mathcal{D}$ are related to those of $\mathcal{R}$ by

$$\rho_1^\mathcal{D}(Q) = \rho_1(Q) - c,$$
$$\rho_2^\mathcal{D} = 2\rho_2 - 2c^2,$$
$$\rho_3^\mathcal{D} = 3\rho_3 - 8c\rho_2 + 5c^3,$$
$$\rho_4^\mathcal{D} = 4\rho_4 - 14c\rho_3 - 6\rho_2^2 + 30c^2\rho_2 - 14c^4,$$

for any P. (These last results are relevant in Chapter 12.)

**Exercise 7.7.** Continue considering the "delicate" theory with $b = 0$ from Exercise 6.1.

(i) Show that the $\mu$ RG equation for $\mathcal{R}$ gives $\mu\partial r_1/\partial\mu = 0$. Thus, as one may also see from the discussion in Sec. 3.6, there are no ultraviolet divergences at next-to-leading order.

(ii)  Show that the RS parameters can be chosen to be

$$g_1 \qquad \text{with } \frac{\partial a}{\partial g_1} = O(a^2),$$

$$\tau \equiv h \, \ln(\mu/\tilde{\Lambda}) \quad \text{with } \frac{\partial a}{\partial \tau} = \frac{\beta(a)}{h} = O(a^3),$$

$$g_3, g_4, \ldots \qquad \text{with } \frac{\partial a}{\partial g_j} = O(a^{j+1}),$$

and that the analogues of the $\rho_n$ invariants, defined via the EC $\beta$ function, are

$$\sigma_1 = g_1 - r_1,$$

$$\sigma_2 = g_2 - 2g_1 r_1 + r_1^2,$$

$$\sigma_3 = g_3 - 3g_2 r_1 + 4g_1 r_1^2 - g_1 r_2 - 2r_1 r_2 + r_3,$$

and so on, together with a $Q$-dependent invariant

$$\sigma_2(Q) = h \, \ln(\mu/\tilde{\Lambda}) + r_2 - r_1^2 - \frac{1}{2} g_1^2.$$

Note that $\sigma_2 - \sigma_1^2 = g$, where $g \equiv g_2 - g_1^2$ is the invariant found in Exercise 6.1. Note also that $g_2$ is not an independent scheme parameter; it is fixed in terms of $g_1$ by $g_2 = \sigma_2 - \sigma_1^2 + g_1^2$.

(iii)  Show that the analogues of the $B_j(a)$ functions, defined now as

$$B_j(a) \equiv \frac{(j-2)}{a^{j+1}} \frac{\partial a}{\partial g_j} = 1 + O(a),$$

are

$$B_1(a) = B(a) \left( 1 - a \int_0^a \frac{dx}{x^2} \left( \frac{(1 + 2g_1 x)}{B(x)^2} - 1 \right) \right),$$

and

$$B_j(a) = \frac{(j-2)}{a^{j-2}} B(a) \int_0^a dx \frac{x^{j-3}}{B(x)^2}, \qquad (j \geq 3),$$

where

$$B(a) \equiv 1 + g_1 a + g_2 a^2 + \cdots.$$

Since $g_2$ is not an independent variable, there is no $B_2(a)$. (In a sense it is $B(a)$.)

# Chapter 8

# Finite Orders and Optimization

## 8.1. Finite-Order Approximants

So far the discussion has been at the level of formal power series. We now need, for the renormalized series case, something corresponding to the notion of the *partial sum* of an ordinary power series. Here *two* truncations are involved, for $\mathcal{R}$ itself and for $\beta$. A definite approximation to $\beta$ is needed because it is the int-$\beta$ equation that relates $a$ back to the theory's one free parameter, the $\tilde{\Lambda}$ parameter of some reference renormalization prescription.

The $(k+1)$th order, or (next-to)$^k$-leading order (N$^k$LO) approximant is naturally defined with both $\mathcal{R}$ and $\beta$ truncated after $k+1$ terms:

$$\mathcal{R}^{(k+1)} \equiv a^{\mathrm{P}}(1 + r_1 a + \cdots + r_k a^k), \qquad (8.1)$$

where $a$ here is shorthand for $a^{(k+1)}$, the solution to the int-$\beta$ equation with $\beta$ replaced by $\beta^{(k+1)}$:

$$\beta^{(k+1)} \equiv -ba^2(1 + ca + \cdots + c_k a^k). \qquad (8.2)$$

It is straightforward to check that the order of the error term $\mathcal{R} - \mathcal{R}^{(k+1)}$ is determined by whichever truncation, of $\mathcal{R}$ or $\beta$, is the more severe. Thus, it is natural to use the same number of terms in each. See Exercises 8.1 and 8.2.

**Other types of approximant** can, of course, be defined: For example, one could use Padé approximant forms for $\mathcal{R}$ and/or for $\beta$. The issue of the RS dependence of such approximants, and how to "optimize" them, would require a fairly straightforward generalization of the following discussion for truncated power series.

## 8.2. Optimization in Low Orders

While the exact $\mathcal{R}$ is RG-invariant, the finite-order approximants are not, since the truncations spoil the cancellations in the RG equations (7.3). If $\mathcal{R}$ in those equations is replaced by $\mathcal{R}^{(k+1)}$ then the right-hand side is not zero but is some remainder term $O(a^{P+k+1})$. The idea of "optimized perturbation theory" is to choose an "optimal" RS in which the approximant $\mathcal{R}^{(k+1)}$ is stationary with respect to RS variations; i.e., the RS in which $\mathcal{R}^{(k+1)}$ satisfies the RG equations, (7.3), with no remainder:

$$\left( \frac{\partial \mathcal{R}^{(k+1)}}{\partial \tau} \bigg|_a + \frac{\beta(a)}{b} \frac{\partial \mathcal{R}^{(k+1)}}{\partial a} \right)_{\text{opt.RS}} = 0, \quad \text{``}j = 1\text{''} \qquad (8.3)$$

$$\left( \frac{\partial \mathcal{R}^{(k+1)}}{\partial c_j} \bigg|_a + \beta_j(a) \frac{\partial \mathcal{R}^{(k+1)}}{\partial a} \right)_{\text{opt.RS}} = 0. \quad j = 2, \dots, k. \quad (8.4)$$

We assume here that the QFT calculations of the $\mathcal{R}$ and $\beta$-function coefficients up to and including $r_k$ and $c_k$ have been done in some calculationally convenient RS. From those results, the values of the invariants $\rho_1(Q)$ and $\rho_1 \equiv c$ and $\rho_2, \dots, \rho_k$ can be obtained. The "optimized" result can be expressed solely in terms of those invariants, and thus has no dependence whatsoever on the choice of RS used for the Feynman-diagram calculations.

**First order** (leading order), as mentioned in Chapter 2, is only a qualitative approximation. It is monotonic in the RS variable $\tau$ and so is not optimizable. There are no invariant quantities, besides $b$. One cannot do better than to guess at a suitable $\mu$ "of order $Q$."

Let us now consider the **second-order** (next-to-leading-order) approximant:

$$\mathcal{R}^{(2)} = a^{\mathrm{P}}(1 + r_1 a), \tag{8.5}$$

$$\beta^{(2)} = -ba^2(1 + ca), \tag{8.6}$$

where $a$ here is short for $a^{(2)}$, the solution to the int-$\beta$ equation (6.24) with $\beta$ replaced by $\beta^{(2)}$:

$$\tau = K^{(2)}(a) = \frac{1}{a} + c\ln\left|\frac{ca}{1+ca}\right|. \tag{8.7}$$

Since $\mathcal{R}^{(2)}$ depends on RS only through the variable $\tau$, only the "$j = 1$" equation (8.3) above is non-trivial. Thus, the optimized $\mathcal{R}^{(2)}$ is determined by a single optimization equation:

$$\frac{\partial r_1}{\partial \tau}\bar{a}^{\mathrm{P}+1} - \bar{a}^2(1 + c\bar{a})(\mathrm{P}\bar{a}^{\mathrm{P}-1} + (\mathrm{P}+1)\bar{r}_1\bar{a}^{\mathrm{P}}) = 0. \tag{8.8}$$

(Overbars are used to indicate the value in the optimum RS.) As discussed in Sec. 7.3, the $a^{\mathrm{P}+1}$ terms must cancel in any RS, which fixes $\frac{\partial r_1}{\partial \tau} = \mathrm{P}$, leaving

$$\mathrm{P} - (1 + c\bar{a})(\mathrm{P} + (\mathrm{P}+1)\bar{r}_1\bar{a}) = 0. \tag{8.9}$$

This equation determines the optimized coefficient $\bar{r}_1$ in terms of the invariant $c$ and the optimized couplant $\bar{a}$:

$$\bar{r}_1 = -\frac{\mathrm{P}}{(\mathrm{P}+1)}\frac{c}{(1+c\bar{a})}. \tag{8.10}$$

But $\bar{r}_1$ is related to $\bar{\tau}$ by the definition of the $\rho_1(Q)$ invariant in Eq. (7.32):

$$\rho_1(Q) \equiv \bar{\tau} - \frac{\bar{r}_1}{\mathrm{P}}. \tag{8.11}$$

Eliminating $\bar{r}_1$ between these last two equations and substituting into the second-order int-$\beta$ equation, (8.7), gives

$$\frac{1}{\bar{a}}\left(1 + c\bar{a}\ln\left|\frac{c\bar{a}}{1+c\bar{a}}\right| + \frac{1}{(\mathrm{P}+1)}\frac{c\bar{a}}{(1+c\bar{a})}\right) = \rho_1(Q). \tag{8.12}$$

If we are interested in a specific $Q$ value then we need to solve this equation numerically for $\bar{a}$ (to a precision comfortably better than our estimated error for $\mathcal{R}^{(2)}$). We can then substitute back in Eq. (8.10) and hence obtain the optimized approximant $\bar{\mathcal{R}}^{(2)} = \bar{a}^{\mathrm{P}}(1 + \bar{r}_1\bar{a})$. However, if we are interested in a range of $Q$ then we may just pick many $\bar{a}$ values, in a suitable range, and evaluate the left-hand side to find the $\rho_1(Q)$, and hence the $Q$ value, corresponding to each of those $\bar{a}$'s.

Note that the only approximations made here are the truncations of the $\mathcal{R}$ and $\beta$ series, leading to Eqs. (8.5), (8.7) that define the second-order approximant is some general RS. We do not, for instance, approximate Eq. (8.10) as $\bar{r}_1 \approx -\frac{\mathrm{P}}{(\mathrm{P}+1)}c$ (which corresponds to the PWMR approximation, discussed later). Nor do we make some uncontrolled analytic approximation to Eq. (8.12). At this order optimization, in practical terms, is no more complicated than the fixed-RS case, where we would have the int-$\beta$ equation to solve.

We now turn to **third order**. The third order approximant is

$$\mathcal{R}^{(3)} = a^{\mathrm{P}}(1 + r_1 a + r_2 a^2), \tag{8.13}$$

where now $a$ is short for $a^{(3)}$, the solution to the int-$\beta$ equation with $\beta$ truncated at third order. $\mathcal{R}^{(3)}$ depends on RS through two parameters $\tau$ and $c_2$, so there are two optimization equations coming from Eqs. (8.3), (8.4). Those equations will mirror Eqs. (7.18), (7.19), generalized to general P. Using the counterparts to Eqs. (7.20), (7.21) they become

$$1 + \left(\frac{(\mathrm{P}+1)}{\mathrm{P}}\bar{r}_1 + c\right)\bar{a} - \bar{B}^{(3)}(\bar{a})\left(1 + \frac{(\mathrm{P}+1)}{\mathrm{P}}\bar{r}_1\bar{a} + \frac{(\mathrm{P}+2)}{\mathrm{P}}\bar{r}_2\bar{a}^2\right) = 0, \tag{8.14}$$

$$1 - \bar{B}_2^{(3)}(\bar{a})\left(1 + \frac{(\mathrm{P}+1)}{\mathrm{P}}\bar{r}_1\bar{a} + \frac{(\mathrm{P}+2)}{\mathrm{P}}\bar{r}_2\bar{a}^2\right) = 0. \tag{8.15}$$

Here $\bar{B}^{(3)}(\bar{a})$ is the $B(a)$ function truncated at third order in the optimum scheme:

$$\bar{B}^{(3)}(\bar{a}) \equiv \left(1 + c\bar{a} + \bar{c}_2\bar{a}^2\right), \tag{8.16}$$

while the other function $\bar{B}_2^{(3)}(\bar{a})$ is obtained from Eqs. (7.9), (7.10) with $B(x)$ replaced by $(1+cx+\bar{c}_2x^2)$. (Note that the only truncations made are the initial truncations of the $\mathcal{R}$ and $B$ series that define the third-order approximant.) The two optimization equations can be combined to yield equations for $\bar{r}_1$ and $\bar{r}_2$ in terms of the integral

$$\bar{I}_2^{(3)} \equiv \int_0^{\bar{a}} \frac{dx}{(1+cx+\bar{c}_2x^2)^2} \tag{8.17}$$

and $\bar{a}$ and $\bar{c}_2$. Combining those equations with the definition of the invariant $\rho_2$ will then determine $\bar{c}_2$ in terms of $\bar{a}$. Finally, $\bar{a}$ itself can be determined by combining the definition of the $\rho_1(Q)$ invariant with the int-$\beta$ equation. In the next chapter, we will explain a systematic method for solving the optimization equations in third and higher orders.

Note that the "optimal RS" is not the same from one order to the next; for instance $\bar{r}_1$ at third order is not the same as $\bar{r}_1$ at second order (so, strictly we should have distinguished $\bar{r}_1^{(2)}$ and $\bar{r}_1^{(3)}$ in the above).

## 8.3. Perturbative Approximants as a Function of the RS Variables

The following discussion is not essential, but the reader may find it helpful to see pictures of the low-order approximants as a function of the RS variables — akin to the figures in Chapters 4 and 5. (For illustrative purposes, we choose the case studied in Example 3 of Chapter 10, but the qualitative features are generic.)

The ($P = 1$) second-order approximant $\mathcal{R}^{(2)} = a(1+r_1a)$, at any given $Q$, is a function of just one RS variable, $\tau$. The coefficient $r_1$ depends linearly on $\tau$ and is $\tau-\rho_1(Q)$, while $a$ is a function of $\tau$ found by inverting the second-order int-$\beta$ equation, Eq. (8.7). Figure 8.1 shows $\mathcal{R}^{(2)}$ as a function of $\tau$. The single maximum corresponds to the optimized result.

One can avoid the inversion step by simply using the int-$\beta$ equation to swap the variable $\tau$ for $a$ itself. (This may be a little mind-boggling at first, but is mathematically quite straightforward.) The approximant $\mathcal{R}^{(2)}$ can then be expressed explicitly as a function

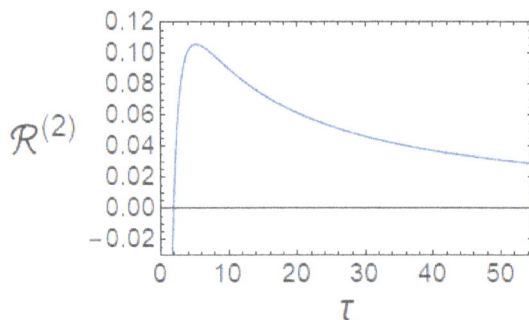

Fig. 8.1.   The second-order approximant as a function of the RS variable $\tau$ in the case corresponding to Example 3 of Chapter 10.

of the new "extraneous parameter" $a$ as

$$\mathcal{R}^{(2)}(a) = a(1 + (K^{(2)}(a) - \rho_1(Q))a)$$

$$= a\left(2 + ca\ln\left(\frac{ca}{1+ca}\right) - \rho_1(Q)a\right). \qquad (8.18)$$

It is nicer to plot this against $1/a$, rather than $a$ itself; the qualitative behaviour is then similar to the previous figure. Figure 8.2 shows a close-up of the region near the maximum. The optimized result corresponds to the maximum of this curve. Thus, in this sense, the optimization condition is $d\mathcal{R}/da = 0$. It is for this reason that we write $\partial\mathcal{R}/\partial a$ for the derivative taken with the $r_i$ coefficients held constant. The total derivative $d\mathcal{R}/da$ here takes into account that the $r_1$ coefficient depends on $\tau$ — and hence, from the current viewpoint, is a function of $a$.

At third order there are two RS variables $\tau$ and $c_2$. It is again mathematically convenient to swap $\tau$ for $a$. Now we must distinguish the usual $\partial\mathcal{R}/\partial a$ from $\partial\mathcal{R}/\partial a|_{c_2}$, which must take into account the $\tau$ dependence, and hence $a$ dependence, of $r_1, r_2$. Similarly, $\partial\mathcal{R}/\partial c_2|_a$ is distinct from the usual $\partial\mathcal{R}/\partial c_2$, which is at constant $\tau$. (See also Exercise 8.3.)

Figure 8.3 shows the third-order approximant $\mathcal{R}^{(3)}$ as a function of $\frac{1}{a}$ and $c_2$. For a fixed $c_2$ value the curves generically have two stationary points or none. However, the optimal result, stationary in

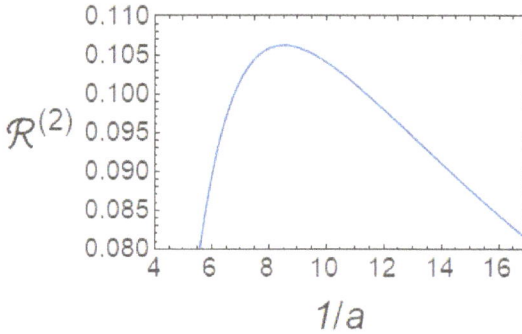

Fig. 8.2. As the previous figure, but with $1/a$ used as the RS variable, and zooming in on the region near the maximum. The optimized result is 0.106 with an error estimate of $\pm 0.011$.

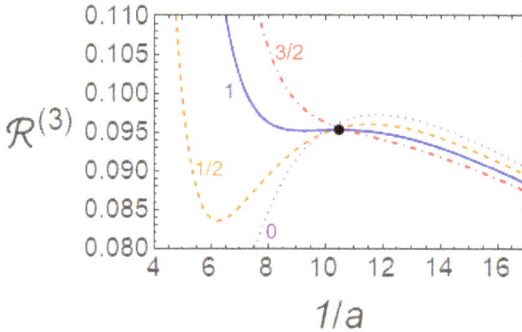

Fig. 8.3. The third-order approximant as a function of the RS variables $1/a, c_2$, in the same case, and shown on the same scale, as the previous figure. The curves are for $c_2 = 0, \frac{1}{2}, 1, \frac{3}{2}$ times the optimal $c_2$ value, which is $-15.9$. The optimal $\frac{1}{a}, c_2$ are indicated by the black dot and give the optimized result $\mathcal{R}^{(3)} = 0.095$, with an error estimate of $\pm 0.005$.

both $1/a$ and $c_2$, is unique and corresponds to a saddle point of the function.

It is instructive to compare this figure with those of Chapter 5. One sees the pattern of diverging, flat, and overdamped regions — and their characteristics show up both going from one order to the next and in the dependence on the other extraneous parameter, $c_2$. Also note that the optimal value of $1/a$ has increased relative to the previous order, in accord with the induced convergence phenomenon.

## 8.4. The Optimization Equations

We now formulate the optimization equations at some general, $(k + 1)$th, order. It is convenient to define

$$S = \frac{1}{\text{P}a^{\text{P}-1}} \frac{\partial \mathcal{R}}{\partial a}, \qquad (8.19)$$

whose series expansion

$$S = 1 + s_1 a + s_2 a^2 + \cdots \qquad (8.20)$$

has coefficients

$$s_m \equiv \left( \frac{\text{P} + m}{\text{P}} \right) r_m. \qquad (8.21)$$

As will be seen below, the use of the $s_m$ coefficients absorbs all the P dependence of the optimization equations. (However, P will reappear later when we need to combine their solution, obtained in the next chapter, with the P-dependent $\rho_n$ invariants.)

Generalizing the discussion in Sec. 7.3 to any P, it must be true that all terms in the RG equations up to and including $O(a^{\text{P}+k})$ cancel automatically in any RS. In the first optimization equation, (8.3), the $\left. \frac{\partial \mathcal{R}^{(k+1)}}{\partial \tau} \right|_a$ term is a polynomial which must cancel the first $k$ terms of $\frac{\beta(a)}{b} \frac{d\mathcal{R}^{(k+1)}}{da}$. A similar observation applies to the other optimization equations, (8.4). Hence, we may reduce the optimization conditions to

$$\bar{B}_j^{(k+1)}(\bar{a}) \bar{S}^{(k+1)}(\bar{a}) - \mathbb{T}_{k-j}[\bar{B}_j^{(k+1)}(\bar{a}) \bar{S}^{(k+1)}(\bar{a})] = 0, \qquad (8.22)$$

for $j = 1, 2, \cdots, k$, where, as in Sec. 5.4, the notation $\mathbb{T}_n[F(a)]$ means "truncate the series for $F(a) = F_0 + F_1 a + \cdots$ immediately after the $a^n$ term" (i.e., $\mathbb{T}_n[F(a)] \equiv F_0 + F_1 a + \cdots + F_n a^n$.)

For future reference, note that the $j = k$ equation, where the $\mathbb{T}_0[\ldots]$ term is just unity, gives

$$\bar{S}^{(k+1)}(\bar{a}) = \frac{1}{\bar{B}_k^{(k+1)}(\bar{a})}. \qquad (8.23)$$

We emphasize again that the $B_j^{(k+1)}(a)$ functions are not polynomials, but are given by Eqs. (7.9), (7.10) with $B(a)$ replaced by $B^{(k+1)}(a) \equiv 1 + ca + \cdots + c_k a^k$.

**Exercise 8.1.** Verify that the $\mathcal{R}$ and $\beta$ series should be truncated after the *same* number of terms to have a formally consistent approximation. As discussed in Sec. 6.4, the int-$\beta$ equation $\tau = K(a)$ can be used to express $a$ as a series in $1/\tau$ (with coefficients involving logarithms of $\tau$). By using $\rho_1(Q) \equiv \tau - r_1/\text{P}$, one can then express $a$, and hence $\mathcal{R}$, as a series in $1/\rho_1(Q)$. Verify that the number of valid terms in the resulting series is controlled by whichever truncation, of $\mathcal{R}$ or $\beta$, is the more severe.

Show explicitly that if two terms of $\beta$ are kept and at least two terms of $\mathcal{R}$ are kept, then

$$\mathcal{R} = \left(\frac{1}{\rho_1}\right)^\text{P} \left(1 - \frac{\text{P} c \ln \rho_1/c}{\rho_1} + O\left(\frac{\text{logs}}{\rho_1^2}\right)\right).$$

Keeping three terms of $\beta$ and $\mathcal{R}$, and noting that $\tilde{\Delta}(a)$ in Eqs. (6.27), (6.28) is $(c^2 - c_2)a + O(a^2)$, the result can be extended to the next order in $1/\rho_1$. For $\text{P} = 1$ the result is

$$\mathcal{R} = \frac{1}{\rho_1}\left(1 - \frac{c \ln \rho_1/c}{\rho_1} + \frac{\rho_2 + c^2 \left(\ln^2 \rho_1/c - \ln \rho_1/c - 1\right)}{\rho_1^2} + O\left(\frac{\text{logs}}{\rho_1^3}\right)\right).$$

Note that the $r_1, r_2, c_2$ coefficients only appear in the invariant combination $\rho_2$.

[This expansion in $1/\rho_1(Q)$ is suitable for the present formal purpose: It is not, however, a solution to the RS-dependence problem. The expansion parameter $1/\rho_1(Q)$, though it is RS invariant, is rather arbitrary because the $\Lambda$ definition is merely a convention.]

**Exercise 8.2.** Show that "mixed-order" approximants, where $\mathcal{R}$ and $\beta$ are truncated at different orders, are not optimizable because they have a monotonic dependence on one or more of the RS parameters $\tau, c_2, c_3, \ldots$ and hence have no stationary point.

**Exercise 8.3.** Show that the optimized result is independent of the particular choice of RS "coordinate system" used: that is, the result for the optimized approximant $\overline{\mathcal{R}}^{(k+1)}$ is the same whether one uses $\tau, c_2, c_3, \ldots$ as scheme parameters or the more general parameterization by $u_1, u_2, \ldots$ as in Exercise 7.2. (This is just the obvious fact that the *value* of a function $f(x_1, x_2, \ldots)$ at a stationary point is invariant under changes of variable $x_1, x_2, \ldots \to x'_1, x'_2, \ldots$, but it is perhaps instructive to check this.) Consider the case of $\mathcal{R}^{(3)}$ explicitly; the generalization to any order will then be obvious. Show first that the optimization equations $\partial \mathcal{R}^{(3)} / \partial u_i$ for $i \geq 3$ are identically satisfied, and that those for $i = 1$ and $i = 2$ are a linear combination of the corresponding $\tau$ and $c_2$ equations.

# Chapter 9

# Solution for the Optimized $r_m$ Coefficients and Optimization Algorithm

## 9.1. Definition of the $H_i(a)$ Functions

In this chapter, it is implicit that all quantities are in the optimal RS at $(k+1)$th order; overbars and $(k+1)$ superscripts will be omitted. Also, we make the convention that

$$r_0 \equiv s_0 \equiv c_0 \equiv 1, \quad \text{and} \quad c_1 \equiv c. \tag{9.1}$$

Next — for reasons that will become clear in the next section — we define some functions $H_1(a), \ldots, H_k(a)$ that are combinations of the $B_1(a), \ldots, B_k(a)$ functions:

$$H_i(a) \equiv \sum_{j=0}^{k-i} c_j a^j \left( \frac{i-j-1}{i+j-1} \right) B_{i+j}(a), \quad i = (1), 2, \ldots, k. \tag{9.2}$$

For $i = 1$ this definition, as it stands, is ambiguous; it should be interpreted as

$$H_1(a) = B_1(a) - \sum_{j=1}^{k-1} c_j a^j B_{j+1}(a), \tag{9.3}$$

corresponding to

$$\lim_{i \to 1} \left( \frac{i-j-1}{i+j-1} \right) = \begin{cases} 1, & j = 0, \\ -1, & j \neq 0. \end{cases} \tag{9.4}$$

Table 9.1.  The $H_i$ functions at low orders. ($H_0 = 1$.)

| $k$ | $H_1$ | $H_2$ | $H_3$ | $H_4$ |
|---|---|---|---|---|
| 1 | $B$ | 0 | | |
| 2 | $B - caB_2$ | $B_2$ | 0 | |
| 3 | $B - caB_2 - c_2a^2B_3$ | $B_2$ | $B_3$ | 0 |
| 4 | $B - caB_2 - c_2a^2B_3 - c_3a^3B_4$ | $B_2 - \frac{1}{3}c_2a^2B_4$ | $B_3 + \frac{1}{3}caB_4$ | $B_4$ |

It is also convenient and natural to define

$$H_0(a) \equiv 1 \quad \text{and} \quad H_{k+1}(a) \equiv 0. \tag{9.5}$$

Note that for the case $i = k$, the definition (9.2) gives $H_k(a) = B_k(a)$. The $H$ functions in low orders are given explicitly in Table 9.1.

In general, the $H$'s are combinations of the $B$'s. It turns out that there is a simple formula for the inverse relationship, giving the $B$'s as combinations of the $H$'s.

**Lemma.**

$$B_j(a) = \sum_{q=0}^{k-j} W_q^j a^q H_{j+q}(a), \quad j = 1, \ldots, k, \tag{9.6}$$

where the $W_i^j$ coefficients are those of the series expansion of $B_j(a)$, Eq. (7.11). (One might describe this result as follows: Take the power series for $B_j(a)$ and truncate it after the $a^{k-j}$ term. Now reweight each term, replacing $a^q$ by $a^q H_{j+q}(a)$, and the result is the full series for $B_j(a)$.)

**Proof.** (We assume $j \neq 1$ for the present.) Using the definition of the $H$'s, Eq. (9.2), the right-hand side becomes

$$\sum_{q=0}^{k-j} W_q^j a^q \sum_{p=0}^{k-j-q} c_p a^p \frac{(j+q-p-1)}{(j+q+p-1)} B_{j+q+p}(a). \tag{9.7}$$

Reorganizing the double sum by defining $n = q + p$ converts this expression to

$$\sum_{n=0}^{k-j} \frac{a^n B_{j+n}(a)}{(j+n-1)} \sum_{p=0}^{n} (n+j-1-2p)c_p W_{n-p}^j. \tag{9.8}$$

The inner sum reduces to $(j - 1)\delta_{n0}$ by virtue of Eq. (7.14). Thus, only the $n = 0$ term of the outer sum survives, the $(j - 1)$ factors cancel, and one is left with just $B_j(a)$, as claimed.

In the $j = 1$ case Eq. (9.6) becomes

$$B_1(a) = \sum_{q=0}^{k-1} c_q a^q H_{q+1}(a). \tag{9.9}$$

Using Eq. (9.3) for $H_1(a)$ and Eq. (9.2) for the other $H$'s, the right-hand side becomes

$$B_1(a) - \sum_{j=1}^{k-1} c_j a^j B_{j+1}(a) + \sum_{q=1}^{k-1} c_q a^q \sum_{j=0}^{k-q-1} c_j a^j \frac{(q-j)}{(q+j)} B_{q+j+1}(a). \tag{9.10}$$

Reorganizing the double sum by defining $n = q + j$ yields

$$B_1(a) - \sum_{j=1}^{k-1} c_j a^j B_{j+1}(a) + \sum_{n=1}^{k-1} \frac{a^n}{n} B_{n+1}(a) \left( \sum_{q=1}^{n} (2q - n) c_q c_{n-q} \right). \tag{9.11}$$

The inner sum, in parentheses, after adding and subtracting a $q = 0$ term becomes

$$n c_n + \sum_{q=0}^{n} (2q - n) c_q c_{n-q}, \tag{9.12}$$

which reduces to $n c_n$ since the summation term vanishes, as noted in Eq. (7.15). Thus, the two series terms in (9.11) cancel leaving just $B_1(a)$, as claimed.    $\square$

## 9.2. Formula for the Optimized $s_m$ Coefficients

We are now ready to state the main result; an exact, analytic expression for the optimized $s_m$, and hence the $r_m = \frac{P}{P+m} s_m$, coefficients, for $m = 0, 1, \ldots, k$, in terms of the (optimized values of) $a$ and the $\beta$-function coefficients $c_2, \ldots, c_k$:

**Theorem.**

$$s_m = \frac{a^{-m}}{B_k(a)} \left( H_{k-m}(a) - H_{k-m+1}(a) \right), \quad m = 0, 1, \ldots, k. \quad (9.13)$$

**Proof.** In the case $j = k$, as noted in Eq. (8.23), the optimization equation reduces to

$$S = \frac{1}{B_k(a)}. \quad (9.14)$$

We first prove that this equation is satisfied. Substituting Eq. (9.13) into the series for $S$ gives

$$S \equiv \sum_{m=0}^{k} s_m a^m = \frac{1}{B_k(a)} \sum_{m=0}^{k} \left( H_{k-m}(a) - H_{k-m+1}(a) \right). \quad (9.15)$$

The $H$'s cancel in pairs leaving

$$S = \frac{1}{B_k(a)} \left( H_0 - H_{k+1} \right) = \frac{1}{B_k(a)}, \quad (9.16)$$

since we defined $H_0 \equiv 1$ and $H_{k+1} \equiv 0$ above.

Using this result and writing out the truncated-series term explicitly, the remaining optimization equations of Eq. (8.22) can be rewritten as

$$\frac{B_j(a)}{B_k(a)} = \sum_{i=0}^{k-j} a^i \sum_{m=0}^{i} s_m W_{i-m}^j \quad j = 1, \ldots, k-1. \quad (9.17)$$

We now need to prove that these equations are satisfied by Eq. (9.13). The right-hand side becomes

$$\sum_{i=0}^{k-j} a^i \sum_{m=0}^{i} W_{i-m}^j \frac{a^{-m}}{B_k(a)} \left( H_{k-m}(a) - H_{k-m+1}(a) \right). \quad (9.18)$$

Reorganizing the double summation, defining $q = i - m$ and thereby replacing $i$ with $m + q$ yields

$$\frac{1}{B_k(a)} \sum_{q=0}^{k-j} a^q W_q^j \sum_{m=0}^{k-j-q} \left( H_{k-m}(a) - H_{k-m+1}(a) \right). \quad (9.19)$$

The inner summation reduces to $H_{j+q}(a)$ since the $H$'s again cancel in pairs (and $H_{k+1} \equiv 0$). Thus, the right-hand side of (9.17) reduces to

$$\frac{1}{B_k(a)} \sum_{q=0}^{k-j} a^q W_q^j H_{j+q}(a) = \frac{B_j(a)}{B_k(a)}, \tag{9.20}$$

where the last step uses the lemma, Eq. (9.6), and produces the left-hand side of (9.17), completing the proof.    □

## 9.3.  An Identity and the PWMR Approximation

It is worth noting the following set of "complete-sum identities:"

$$\sum_{j=0}^{k} c_j a^j \left( \frac{i - j - 1}{i + j - 1} \right) B_{i+j}(a) = 1, \quad i = (1), 2, \ldots, k, \tag{9.21}$$

with the $i = 1$ case interpreted using (9.4). The proof is given in Appendix 9.A, which discusses various properties of the $I_j(a)$ integrals related to the $B_j(a)$ functions.

These identities reveal a remarkable property of the $H_i(a)$'s, which are defined as a "partial sum" (over $j = 0, \ldots, k - i$) of the same terms. Hence, we can write

$$H_i(a) = 1 - \sum_{j=k-i+1}^{k} c_j a^j \left( \frac{i - j - 1}{i + j - 1} \right) B_{i+j}(a), \quad i = (1), 2, \ldots, k,$$
$$\tag{9.22}$$

which, unlike the $H_i$ definition, involves $B_j$'s with $j$ greater than $k$. Since the $B_j$'s all start $1 + O(a)$ we see that the series for $H_i(a) - 1$ begins only at order $a^{k-i+1}$:

$$H_i(a) - 1 = \frac{k - 2i + 2}{k} c_{k-i+1} a^{k-i+1} (1 + O(a)). \tag{9.23}$$

Substituting this result into Eq. (9.13) leads to

$$s_m = a^{-m} \left( (1 + O(a^{m+1})) - \left( 1 + \frac{(-k + 2m)}{k} c_m a^m \right) \right) \frac{1}{1 + O(a)} \tag{9.24}$$

and hence

$$s_m = \frac{k - 2m}{k} c_m + O(a).$$
(9.25)

This result was first obtained — in a quite different manner — by Pennington, Wrigley, and Mignaco and Roditi (PWMR). The resulting PWMR approximation (see Exercise 9.1) can be useful when $a \ll 1$ (one also needs $a \ll a^*$ if a finite infrared limit $a^*$ exists) and it provides a good starting point for the full optimization procedure. (Obviously, post-PWMR, post-post-PWMR, ... approximations can be obtained by expanding further in powers of $a$; see Exercise 9.3.)

## 9.4. Results in Terms of the $I_j$ Integrals

Recalling that the $B_j(a)$ functions are related to the $I_j$'s by Eq. (7.9), we may rewrite the $H$'s as sums of $I$'s rather than $B$'s:

$$H_i(a) = \frac{B(a)}{a^{i-1}} \sum_{j=0}^{k-i} (i - j - 1) c_j I_{i+j}, \quad i = 2, \dots, k,$$
(9.26)

and

$$H_1(a) = B(a) \left( 1 - \sum_{j=1}^{k-1} j c_j I_{j+1} \right).$$
(9.27)

Note that at $(k+1)$th order only $I_2, \dots, I_k$ arise. The $s_m$ coefficients, from Eq. (9.13), can thus be expressed in terms of the $I_j$'s (see Exercise 9.3). The result, defining $c_{-1} \equiv 0$, is

$$s_m = \frac{1}{(k-1)I_k} \sum_{j=0}^{m} I_{k-m+j} \Big( (k - m - j - 1) c_j$$
$$- (k - m - j + 1) \frac{c_{j-1}}{a} \Big),$$
(9.28)

for $m = 1, \dots, k - 2$. The cases $m = k - 1, k$, involving $H_1, H_0$, need special treatment. Those results are

$$s_{k-1} = \frac{1}{(k-1)I_k} \left( 1 - \sum_{j=1}^{k-1} I_{j+1} \left( j c_j - (j - 2) \frac{c_{j-1}}{a} \right) \right)$$
(9.29)

Table 9.2.   The $s_m$ coefficients at low orders, in the terms of the $I_j$ integrals. $B \equiv \sum_{j=0}^{k} c_j a^j$.

| $k$ | $s_m$ coefficients |
|---|---|
| 1 | $s_1 = -\frac{c}{B}$. |
| 2 | $s_1 = -c - \frac{1}{a} + \frac{1}{I_2}$,                     $s_2 = \frac{c}{a} + \frac{1}{aI_2}\left(\frac{1}{B} - 1\right)$. |
| 3 | $s_1 = -\frac{1}{a} + \frac{I_2}{2I_3}$,                  $s_2 = -c_2 + \frac{1}{2aI_3}\left(a - (1 + ca)I_2\right)$, |
| | $s_3 = \frac{c_2}{a} + \frac{1}{2aI_3}\left(\frac{1}{B} - 1 + cI_2\right)$. |
| 4 | $s_1 = \frac{c}{3} - \frac{1}{a} + \frac{2I_3}{3I_4}$,           $s_2 = -\frac{(c+c_2 a)}{3a} + \frac{1}{3aI_4}\left(aI_2 - 2I_3\right)$, |
| | $s_3 = \frac{(c_2 - 3c_3 a)}{3a} + \frac{1}{3aI_4}$          $s_4 = \frac{c_3}{a} + \frac{1}{3aI_4}\left(\frac{1}{B} - 1 + cI_2 + 2c_2 I_3\right)$. |
| | $\left(a - (1 + ca)I_2 - 2c_2 aI_3\right)$, |

and

$$s_k = \frac{1}{(k-1)aI_k}\left(\frac{1}{B} - 1 + \sum_{j=1}^{k-1} jc_j I_{j+1}\right). \qquad (9.30)$$

For low orders, the results are collected in Table 9.2. The $I_j$ integrals are readily evaluated by computer algebra; see Table 9.3. Properties of the $I$'s are discussed in Appendix 9.A.

## 9.5. Optimization Algorithm

The optimization problem at $(k+1)$th order involves $2k+1$ variables, namely, $a, \tau, c_2, \ldots, c_k$, and $r_1, \ldots, r_k$. These are connected by $2k+1$ equations, namely, the int-$\beta$ equation, the $k$ optimization equations, and the $k$ formulas for the invariants $\rho_1(Q)$ and $\rho_2, \ldots, \rho_k$ (whose numerical values we assume are given — in the case of $\rho_1(Q)$ the numerical value will depend on the value of $Q$ being considered). We shall use $a, c_2, \ldots, c_k$ as the principal variables. The solution to the optimization equations just discussed gives the coefficients $r_1, \ldots, r_k$ directly in terms of the principal variables. The int-$\beta$ equation explicitly fixes $\tau$ in terms of the principal variables. The $\rho_1(Q) = \tau - r_1$ equation can be used at the end to relate $a$ to $Q$, so the remaining task is to use the formulas for $\rho_2, \ldots, \rho_k$ to determine, self-consistently, by some convergent iterative procedure,

Table 9.3.   Mathematica notebook to evaluate the $I_2, I_3$, and $\tilde{\Delta}$ integrals for the $k = 3$ case. The output expressions are not shown for reasons of space. (To avoid **ConditionalExpression** forms, it is tacitly assumed that $a$ is less than the smallest positive root of $B(x)$, as must be true in the present context.)

---

*In[#]*  **\$Assumptions = {c ∈ Reals, c2 ∈ Reals, c3 ∈ Reals, a > 0};**
*In[#]*  **B = 1 + c x + c2 x^2 + c3 x^3;**
*In[#]*  **int2 = Integrate $\left[\frac{1}{B^2}, x\right]$ ;**
*In[#]*  **Ii2 = (int2 /. x → a) − (int2 /. x → 0)**
*Out[#]*  ---
*In[#]*  **int3 = Integrate $\left[\frac{x}{B^2}, x\right]$ ;**
*In[#]*  **Ii3 = (int3 /. x → a) − (int3 /. x → 0)**
*Out[#]*  ---
*In[#]*  **Deltwint = Integrate[(1/x^2)(1/B − 1 + c x), x];**
*In[#]*  **Deltw = (Deltwint /. x → a) − (Deltwint /. x → 0)**
*Out[#]*  ---

---

the $c_2, \ldots, c_k$ variables. One such algorithm is the following (recall $s_m \equiv \left(\frac{P+m}{P}\right) r_m$):

**(1)** Choose a numerical value for $a$.
**(2)** Make an initial guess for the numerical values of $s_1, \ldots, s_k$.
**(3)** Find values for the $c_j$'s from the invariants using Eq. (7.33).
**(4)** Obtain new values for the $s_m$ from the formulas of the preceding section.
**(5)** Iterate from step **3** until the results converge to the desired precision.
**(6)** Finally, use

$$\rho_1(Q) = \frac{1}{a} + c \ln|ca| - \tilde{\Delta}(a) - \left(\frac{P}{P+1}\right) s_1 \qquad (9.31)$$

(from the definition of $\rho_1(Q)$ as $\tau - r_1$ combined with the int-$\beta$ equation $\tau = \hat{K}(a)$ in the form of Eq. (6.27)) to find the value of $Q$ that corresponds to the chosen $a$ value. One can then repeat the whole procedure with different initial $a$ values to cover the desired range of $Q$ values — or to home in on one particular $Q$ value.

It is usually convenient to start at a very small $a$ value and use the PWMR approximation (see Exercise 9.1) for the initial $s_m$ values. One can then move step-by-step to larger $a$ values (and hence lower $Q$'s) using the previous $s_m$'s as the initial guess for the next step.

To illustrate the algorithm we show, in Table 9.4, a Mathematica notebook for the fourth-order ($k = 3$, P $= 1$) case. The desired values of $c, \rho_2, \rho_3$ and $a$ should be entered in place of those used there (which correspond to Example 3 in Chapter 10).

Various details of the algorithm can be refined. It appears to be quite robust and efficient except in the far-infrared region, where more elaborate numerical analysis techniques may be needed. The infrared limit can be analyzed analytically, as will be discussed in Chapter 11.

> **In the case of a fixed-point limit,** where the $\beta$ function has a zero at $a = a^*$, one can avoid the worst problems at low $Q$ by iterating at a fixed $B(a)$ value rather than at a fixed $a$ value. That is, after step **3**, one constructs the new $B(x)$ function and solves for a new $a$ from $B(a) = B_0$, where $B_0$ is the value obtained in the first iteration.

> **Is the optimal solution determined uniquely?** This seems very likely, due to the fact that $k$ extraneous parameters are involved at $(k + 1)$th order. A general proof appears difficult, but at large $Q$, where the PWMR approximation is good, the uniqueness can be proved easily: The PWMR $c_j$ coefficients are fixed directly by the values of the $\rho_n$ invariants (see Exercise 9.1), leaving Eq. (9.31) to fix the optimal $a$. The right-hand side is easily seen to be a monotonic-decreasing function of $a$, while the left-hand side is $a$-independent; hence, their intersection point is unique.

## Appendix 9.A: Properties of the $I_j$ and $J_j$ Integrals

In this appendix, we discuss some properties of the integrals

$$I_j(a) \equiv \int_0^a dx \frac{x^{j-2}}{B(x)^2}, \quad J_j(a) \equiv \int_0^a dx \frac{x^{j-2}}{B(x)}. \qquad (9A.1)$$

It should be understood that we are considering $(k + 1)$th order, so that "$B(x)$" is really shorthand for $B^{(k+1)}(x) \equiv 1 + cx + \cdots + c_k x^k$. Unless otherwise stated, we assume $j \geq 2$, so that the integrals

Table 9.4. A basic program to implement the optimization algorithm for $P = 1$ at fourth order ($k = 3$). Expressions for $I_2, I_3, \tilde{\Delta}$ should be cut-and-paste into the opening lines from the outputs of Table 9.3.

---

*In[#]*    Ii2 = ...; Ii3 = ...; Deltw = ...;

*In[#]*    c = 115/58; $\rho$2 = −9.92498; $\rho$3 = −115.21021;

*In[#]*    c2frinv = $\rho$2 − $\left(\text{r2} - \text{c r1} - \text{r1}^2\right)$ /. {r1 → s1/2, r2 → s2/3};
      c3frinv = $\rho$3 − $\left(-2\,\text{c2 r1} + \text{c r1}^2 + 4\,\text{r1}^3 - 6\,\text{r1 r2} + 2\,\text{r3}\right)$ /.
      {r1 → s1/2, r2 → s2/3, r3 → s3/4};

*In[#]*    reslist = {};

*In[#]*    (* Initial guesses for s1, s2, s3. PWMR used here. *)
      s1 = $\frac{c}{3}$; s2 = $-\frac{3}{8}\left(\rho 2 + \frac{7}{36}\text{c}^2\right)$;
      s3 = $-2\left(\rho 3 + \frac{1}{4}\rho 2\,\text{c} + \frac{1}{432}\text{c}^3\right)$;

*In[#]*    (*%%%%*)

*In[#]*    aa = 0.0899359; n = 0;

*In[#]*    (*####*)
      n = n + 1;
      c2 = c2frinv; c3 = c3frinv;
      Bval = 1 + c a + c2 a^2 + c3 a^3 /. a → aa;
      Ii2val = Re[Ii2 /. a → aa];
      Ii3val = Re[Ii3 /. a → aa];
      s1 = −1/aa + Ii2val/(2 Ii3val);
      s2 = −c2 + (aa − (1 + c aa) Ii2val)/(2 aa Ii3val);
      s3 = c2/aa + (1/Bval − 1 + c Ii2val)/(2 aa Ii3val);
      RR[n] = a(1 + (s1/2)a + (s2/3)a^2 + (s3/4)a^3) /. a → aa;
      (*####*)

*In[#]*    (* Evaluate the preceding cell several times until satisfactorily converged.
      Use ListPlot[Table[RR[i], {i, 1, n}]] to view convergence. *)

*In[#]*    rho1Q = (1/a + c Log [Abs [c a]] − Re [Deltw] − s1/2) /. a → aa;

*In[#]*    AppendTo [reslist, {aa, rho1Q, RR[n], c2, c3, s1, s2, s3}]

*Out[#]*   {{0.0899359, 6.02383, 0.0913479, −7.42689, −230.263,
      −0.125903, −7.85684, 233.94}}

*In[#]*    (*Return to (*%%%%*) and enter a somewhat larger aa value,
      thereby moving to lower Q. *)

---

are convergent. These integrals may be evaluated analytically by expressing the polynomial $B(x)$ as a product of its factors and using partial fractions. (The resulting expressions are cumbersome, and complicated by the possible presence of complex roots in pairs, but are readily handled by computer algebra programs. Note that, in the present context, $a$ will naturally always be smaller than any positive root of $B(x)$.)

The $J$'s are given by a sum of $I$'s:

$$J_i = \sum_{j=0}^{k} c_j \, I_{i+j}. \tag{9A.2}$$

The proof is simple: The right-hand side is

$$\sum_{j=0}^{k} c_j \int_0^a dx \frac{x^{i+j-2}}{B(x)^2} = \int_0^a dx \frac{x^{i-2}}{B(x)^2} \sum_{j=0}^{k} c_j x^j, \tag{9A.3}$$

and the sum gives $B(x)$, which cancels with one of the $B(x)$ factors in the denominator, so that the integral reduces to $J_i$.

A set of "complete-sum identities" follows from the fact that

$$\int_0^a dx \frac{d}{dx} \left( \frac{x^{i-1}}{B(x)} \right) = \frac{a^{i-1}}{B(a)}. \tag{9A.4}$$

The left-hand side is

$$\int_0^a dx \left( (i-1)\frac{x^{i-2}}{B(x)} - \frac{x^{i-1}}{B(x)^2}\frac{dB}{dx} \right) = (i-1)J_i - \sum_{j=0}^{k} jc_j \int_0^a dx \frac{x^{i+j-2}}{B(x)^2}$$

$$= (i-1)\sum_{j=0}^{k} c_j I_{i+j} - \sum_{j=0}^{k} jc_j I_{i+j}$$

$$= \sum_{j=0}^{k} (i-j-1)c_j I_{i+j}. \tag{9A.5}$$

Thus, we obtain a set of complete-sum identities:

$$\sum_{j=0}^{k} (i-j-1)c_j I_{i+j} = \frac{a^{i-1}}{B(a)}, \quad i = (1), 2, 3, \ldots. \tag{9A.6}$$

In the special case $i = 1$, the result should more properly be written as

$$\sum_{j=1}^{k}(-j)c_j I_{j+1} = \frac{1}{B(a)} - 1, \qquad (9A.7)$$

where the sum starts with $j = 1$. (As a mnemonic, one can regard the $j = 0$ term, which has a vanishing coefficient times a divergent $I_1$, as producing a 1 that is taken to the right-hand side.)

Recalling that the $B_j(a)$ functions are related to the $I_j$'s, we can rewrite the complete-sum identities, Eq. (9A.6), in terms of $B_j$'s to get

$$\sum_{j=0}^{k}(i - j - 1)c_j \frac{a^{i+j-1}}{(i+j-1)} \frac{B_{i+j}(a)}{B(a)} = \frac{a^{i-1}}{B(a)}, \qquad (9A.8)$$

which, when divided by $a^{i-1}/B(a)$, gives the result in Eq. (9.21). (The $i = 1$ case needs special consideration, but can be easily checked.)

In the factorization-scheme-dependence problem, discussed in Chapter 12, we also encounter the integrals $J_{k+1}$ and $I_{2k+1}$ at $(k+1)$th order. These can also be reduced to combinations of $I_2, \ldots, I_k$ by the earlier formulas plus the identity

$$\sum_{j=1}^{k} jc_j J_{j+1} = \ln B(a), \qquad (9A.9)$$

which follows from

$$\int_0^a dx \frac{1}{B(x)} \frac{dB(x)}{dx} = \int_0^a dx \frac{1}{B(x)} \sum_{j=0}^{k} jc_j x^{j-1} = \sum_{j=1}^{k} jc_j \int_0^a dx \frac{x^{j-1}}{B(x)}.$$

$$(9A.10)$$

For cases involving $I_j$ or $J_j$ for $j = 1$ or 0, we may define "regulated" versions of the integrals by subtracting off the inverse powers of $x$ in a Laurent expansion of the integrand. However, the only such case we really need mention is

$$\tilde{\Delta} \equiv J_0^{\text{reg}} \equiv \int_0^a dx \left( \frac{1}{x^2 B(x)} - \frac{1 - cx}{x^2} \right), \qquad (9\text{A}.11)$$

which is the integral occurring in the int-$\beta$ equation (see Eqs. (6.27) and (6.28).) In practice, this integral is best evaluated directly, but it is noteworthy that it can be expressed as a sum over convergent $J$ integrals:

$$\tilde{\Delta} = \int_0^a dx \frac{1}{x^2 B(x)} \left( 1 - (1 - cx) \sum_{i=0}^k c_i x^i \right)$$

$$= \int_0^a dx \frac{1}{x^2 B(x)} \left( \sum_{i=2}^k (cc_{i-1} - c_i) x^i + cc_k x^{k+1} \right)$$

$$= \sum_{j=1}^{k-1} (cc_j - c_{j+1}) J_{j+1} + cc_k J_{k+1}. \qquad (9\text{A}.12)$$

The last term, by using the $J$ complete-sum identity, Eq. (9A.9), can be expressed as

$$cc_k J_{k+1} = \frac{c}{k} \left( \ln B(a) - \sum_{j=1}^{k-1} jc_j J_{j+1} \right), \qquad (9\text{A}.13)$$

so that

$$\tilde{\Delta} = \frac{c}{k} \ln B(a) + \sum_{j=1}^{k-1} \left( \frac{(k-j)}{k} cc_j - c_{j+1} \right) J_{j+1}. \qquad (9\text{A}.14)$$

Note that the $J$'s involved can be expressed, using Eq. (9A.2), as a sum of $I_j$'s with $j = 2, \ldots 2k$. Then, by using the complete-sum identities to substitute for $I_{k+1}, \ldots, I_{2k}$, one can reduce the result to a sum over just $I_2, \ldots, I_k$. (See Exercise 9.4.)

**Exercise 9.1.** Combine the PWMR result, Eq. (9.25), with the formulas for the $\rho_i$ invariants to solve for the PWMR $r_m$ coefficients in terms of the invariants for second, third, and fourth orders.

For $\text{P} = 1$ you should find $r_1 = -\frac{1}{2}c$ at second order; $r_1 = 0$, $r_2 = -\frac{1}{2}\rho_2$ at third order; and $r_1 = \frac{1}{6}c$, $r_2 = -\frac{1}{8}\left(\rho_2 + \frac{7}{36}c^2\right)$, $r_3 = -\frac{1}{2}\left(\rho_3 + \frac{1}{4}\rho_2 c + \frac{1}{432}c^3\right)$ at fourth order.

**Exercise 9.2.** Express the formula for the $s_m$ coefficients, Eq. (9.13), in matrix form, with the column vector of $\hat{s}_m$'s (with $\hat{s}_m \equiv s_m a^m$ and $\hat{s}_0 \equiv 1$) given by $1/B_k$ times a matrix of 0's and 1's times a column vector of $H_i$'s. Then find the matrix inverse. Substitute in the lemma, Eq. (9.6), to show that the column vector of $(1, B, B_2, \ldots, B_k)$ is given by the matrix

$$
B_k
\begin{pmatrix}
1 & 1 & 1 & \cdots & 1 & 1 \\
\mathbb{T}_{k-1}(B) & \mathbb{T}_{k-2}(B) & \mathbb{T}_{k-3}(B) & \cdots & 1 & 0 \\
\vdots & \vdots & \vdots & \cdots & \vdots & \vdots \\
\mathbb{T}_2(B_{k-2}) & \mathbb{T}_1(B_{k-2}) & 1 & \cdots & 0 & 0 \\
\mathbb{T}_1(B_{k-1}) & 1 & 0 & \cdots & 0 & 0 \\
1 & 0 & 0 & \cdots & 0 & 0
\end{pmatrix}
$$

times the column vector $(1, \hat{s}_1, \hat{s}_2, \ldots, \hat{s}_k)$.

**Exercise 9.3.**

(i) Derive Eq. (9.28).
(ii) Obtain Eqs. (9.29, 9.30) for the special cases $s_k$ and $s_{k-1}$.
(iii) Use the complete-sum identities (9A.6), to obtain the formula

$$
s_m = \frac{1}{(k-1)I_k}\left((k-2m)c_m\frac{I_{k+1}}{a} + \sum_{j=m+1}^{k} c_j\right.
$$
$$
\left. \times \left((k-m-j)\frac{I_{k-m+j+1}}{a} - (k-m-j-1)I_{k-m+j}\right)\right),
$$

which gives compact results for $s_k, s_{k-1}, \ldots$, but is more cumbersome for $s_1, s_2, \ldots$. Note that it involves integrals beyond $I_k$, up to $I_{2k-m+1}$ (and so up to $I_{2k}$ for $m = 1$). The first term directly yields the PWMR approximation.
(iv) Use the series expansion of the $I_j$'s to obtain the post-PWMR approximation

$$
s_m = \frac{(k-2m)}{k}c_m\left(1 - \frac{2ca}{k(k+1)}\right) + \frac{2(m+1)}{k(k+1)}c_{m+1}a.
$$

**Exercise 9.4.** From Eqs. (9A.14) and (9A.2) show that for $k = 2$

$$\tilde{\Delta} = \frac{c}{2} \ln B(a) + \left( \frac{c^2}{2} - c_2 \right) (I_2 + cI_3 + c_2 I_4).$$

Using the complete-sum identities of Eq. (9A.6), eliminate $I_3$ and $I_4$ and hence express $\tilde{\Delta}$ in terms of $I_2$:

$$\tilde{\Delta} = \frac{c}{2} \ln B(a) - \frac{(c^2 - 2c_2)}{4c_2} \left( (c^2 - 4c_2)I_2 - \frac{a}{B(a)} \left( c^2 - 2c_2 + cc_2 a \right) \right).$$

# Chapter 10

# Numerical Examples for $R_{e^+e^-}$ in QCD

## 10.1. $R_{e^+e^-}$ and Its QCD Corrections

We turn next to illustrative numerical results for a specific physical quantity, namely the QCD corrections to the ratio

$$R_{e^+e^-} \equiv \frac{\sigma_{tot}(e^+e^- \to \text{hadrons})}{\sigma(e^+e^- \to \mu^+\mu^-)}. \tag{10.1}$$

In the parton model, where quarks are treated as free particles, the cross section to produce a quark–antiquark pair, of a specific flavour, is exactly the same as that for production of a $\mu^+\mu^-$ pair, except for a factor of the quark charge squared, $q_i^2$, and a factor of 3, since each flavour quark comes in three colours. It is assumed that the quarks will later "fragment" into hadrons. How that happens does not matter; it happens with probability one, and does not affect the cross section, which is determined by the probability of producing the $q\bar{q}$ pair in the first place. Thus, neglecting masses, the parton model prediction is just

$$R_{e^+e^-}^{\text{parton model}} = 3 \sum_i q_i^2, \tag{10.2}$$

where the sum is over all flavours.

Quarks, however, are not free; they may radiate gluons and exchange virtual gluons, giving rise to QCD corrections. The leading-order correction has a coefficient of unity times the QCD couplant, $a$, so we may write

$$R_{e^+e^-} = \left(3\sum_i q_i^2\right)(1 + \mathcal{R}_{e^+e^-}),\qquad(10.3)$$

where $\mathcal{R}_{e^+e^-}$ is a normalized physical quantity whose QCD perturbation series has the form

$$\mathcal{R}_{e^+e^-} = a(1 + r_1 a + r_2 a^2 + \cdots).\qquad(10.4)$$

It depends upon the energy scale $Q$ by the magic of Dimensional Transmutation, as discussed in Chapter 2. (In this specific context, we choose to define "$Q$" as the total $e^+e^-$ centre-of-mass energy.) The coefficients $r_1$, $r_2$, and $r_3$ have been calculated, as have the $\beta$-function coefficients $c_2$, $c_3$ (all in the $\overline{MS}$ scheme), and are quoted in Appendix 10.A. Thus, we are able to obtain second-, third-, and fourth-order approximants. Our focus here will not be on comparison with experimental data (which would require discussion of other issues) but on the apparent convergence, or otherwise, of results from one order to the next.

> **In the real world,** quarks have masses. Calculating QCD radiative corrections including quark masses is much harder, so generally quark masses are neglected. This means that one is approximating "real QCD" with a *set* of effective theories, each with a different number of massless quarks. The number of "active quarks," $n_f$, depends on the physical energy scale $Q$ being considered. Near a flavour threshold, at $Q = 2m_i$, the quark mass needs to be allowed for, at least kinematically. For $R_{e^+e^-}$, the result including masses at first order (where $\mathcal{R} \approx a$) is
>
> $$R_{e^+e^-} = 3\sum_i q_i^2\, T(v_i)\,(1 + g(v_i)\mathcal{R}),$$
>
> where
>
> $$v_i = \sqrt{1 - 4m_i^2/Q^2},$$
>
> $$T(v) = v(3 - v^2)/2,$$
>
> $$g(v) = \frac{4\pi}{3}\left[\frac{\pi}{2v} - \left(\frac{3+v}{4}\right)\left(\frac{\pi}{2} - \frac{3}{4\pi}\right)\right].$$

(The formula for $g(v)$ is a simplified approximation to the actual, rather cumbersome, formula.) At higher orders, we may use the above form but with the $\mathcal{R}$ calculated to higher orders in the massless effective theory with the appropriate $n_f$. The $\tilde{\Lambda}_{\overline{\text{MS}}}$ parameters of the different effective theories then need to be matched at the flavour threshold so that $\mathcal{R}$ is continuous. For a fuller discussion of the phenomenological issues, see Ref. [15].

## 10.2. Procedure

We shall compare the optimized results with the conventional approach, which is to use the "modified minimal subtraction" ($\overline{\text{MS}}$) prescription with the renormalization scale $\mu$ chosen equal to the centre-of-mass energy, $Q$. (Properly speaking, then, the RS is "$\overline{\text{MS}}(\mu = Q)$.") We will essentially presume that the value of $\tilde{\Lambda}_{\overline{\text{MS}}}$ is known from fitting other experimental data. However, to avoid committing to any specific value, we label our examples, not by $Q$, but by the ratio of $Q$ to $\tilde{\Lambda}_{\overline{\text{MS}}}$. At each order we proceed as if only the coefficients to that order had been calculated.

To obtain the $\overline{\text{MS}}$ results, at a given $Q/\tilde{\Lambda}_{\overline{\text{MS}}}$ value, the first step is to evaluate the numerical value of the $\tau$ parameter of the $\overline{\text{MS}}(\mu = Q)$ scheme:

$$\tau^{\overline{\text{MS}}} = b\ln(Q/\tilde{\Lambda}_{\overline{\text{MS}}}). \tag{10.5}$$

One must then numerically solve the int-$\beta$ equation. At second order, where $B(a)$ is approximated by $1 + ca$, that equation is $\tau^{\overline{\text{MS}}} = K^{(2)}(a)$, with $K^{(2)}(a)$ given by Eq. (6.23). With the resulting $a$, one then evaluates $\mathcal{R}^{(2)}_{\overline{\text{MS}}} = a(1 + r_1^{\overline{\text{MS}}} a)$. At third order one must numerically solve

$$\tau^{\overline{\text{MS}}} = K^{(3)}_{\overline{\text{MS}}}(a) \equiv \frac{1}{a} + c\ln|ca| - \tilde{\Delta}^{(3)}_{\overline{\text{MS}}}(a), \tag{10.6}$$

where $\tilde{\Delta}(a)$ is given by Eq. (6.28) with, in this case, $B(x)$ approximated by $1 + cx + c_2 x^2$ with $c_2 = c_2^{\overline{\text{MS}}}$. With the resulting $a$ one then evaluates $\mathcal{R}^{(3)}_{\overline{\text{MS}}} = a(1 + r_1^{\overline{\text{MS}}} a + r_2^{\overline{\text{MS}}} a^2)$. At fourth order, the procedure is the same, except that one now includes a $c_3^{\overline{\text{MS}}}$ term in $B(x)$ and an $r_3^{\overline{\text{MS}}}$ term in $\mathcal{R}$.

To obtain the optimized results, one first needs to calculate the numerical values of the invariants. At a given $Q/\tilde{\Lambda}_{\overline{MS}}$, one finds $\tau^{\overline{MS}}$ from Eq. (10.5) and then obtains $\rho_1(Q)$ as $\tau^{\overline{MS}} - r_1^{\overline{MS}}$. The $\rho_2$ and $\rho_3$ invariants, which are $Q$-independent, are similarly obtained by evaluating their definitions in Eq. (7.31) using the $\overline{MS}$-scheme $r_i$ and $c_j$ coefficients: The $\rho_2, \rho_3$ results are quoted in Appendix 10.A. The optimized result to second order is obtained from Eqs. (8.12) and (8.10) with $P = 1$. At higher orders, one can use the algorithm described in the preceding chapter.

At each order one wants, not only a result for $\mathcal{R}$ but also an estimate for its likely error. There is no rigorous way of doing this. However, it is reasonable to expect that the "apparent convergence" of the series (i.e., the behaviour of the terms that have been calculated) is some sort of guide. We shall adopt the common practice when dealing with asymptotic series of viewing the magnitude of the last calculated term, $|r_k a^{k+1}|$, as the error estimate. We do this both for the $\overline{MS}$ and the optimized results. The change in the $\mathcal{R}$ results from one order to the next — which for optimization is not the same thing — provides another indication of the likely error; it seems quite consistent with our error estimate.

We will give two sets of examples; one set at moderately high energies, and the other at low energies. For the first set of examples the phenomenologically appropriate number of flavours is $n_f = 5$ $(u, d, s, c, b$ quarks), while for the second set it is $n_f = 2$ $(u, d$ quarks only). Results for $(k + 1)$th order $(k = 1, 2, 3)$ in both the $\overline{MS}$ and optimized schemes are presented in the tables and figures below.

## 10.3. High-Energy Examples

For $n_f = 5$ the $\beta$-function's leading, RS-invariant coefficients are

$$b = \frac{23}{6}, \qquad c = \frac{29}{23}. \tag{10.7}$$

In the $\overline{MS}$ scheme, its next two coefficients are

$$c_2^{\overline{MS}} = \frac{9769}{6624} = 1.474789, \tag{10.8}$$

$$c_3^{\overline{\mathrm{MS}}} = -\frac{26017}{31104} + \frac{11027}{1242}\zeta_3 = 9.835916, \qquad (10.9)$$

where $\zeta_s$ is the Riemann zeta function. The $\overline{\mathrm{MS}}$ coefficients in $\mathcal{R}(e^+e^-)$ are

$$r_1^{\overline{\mathrm{MS}}} = 1.409230, \qquad r_2^{\overline{\mathrm{MS}}} = -12.80463, \qquad r_3^{\overline{\mathrm{MS}}} = -80.43373.$$
$$(10.10)$$

(The exact values, involving $\zeta_2$, $\zeta_3$, $\zeta_5$ and $\zeta_7$, see Appendix 10.A, were used in our calculations.) Inserting these values in Eq. (7.31) yields

$$\rho_2 = -15.0926, \qquad \rho_3 = -33.2216. \qquad (10.11)$$

Our first two examples are cases that have been discussed previously in the literature.

**The history** of these two examples is interesting. Back in 1988, after publication of a result for $r_2^{\overline{\mathrm{MS}}}$, it had seemed that "optimization" gave very unsatisfactory results; the optimized couplant increased from second to third order, the apparent convergence was poor, and the error estimate was large; worse than $\overline{\mathrm{MS}}$. It turned out, though, that the original calculation of $r_2^{\overline{\mathrm{MS}}}$ was incorrect. When the correct result was published in 1991, the situation was transformed. This unfortunate history does at least illustrate the fact that the improvement provided by "optimization" is not trivial or accidental. The more recent result for $r_3^{\overline{\mathrm{MS}}}$ provides further confirmation that optimization works as advertised.

At these energies, the perturbation series seems well behaved. The $\overline{\mathrm{MS}}$ results are quite satisfactory and optimization provides only a slight improvement. It should be borne in mind, though, that the popularity of the $\overline{\mathrm{MS}}$ scheme — over the original minimal-subtraction scheme, for instance — was in part due to it giving sensible-looking results for $R_{e^+e^-}$ when $\mu$ equals the centre-of-mass energy. Applying $\overline{\mathrm{MS}}$ to other quantities requires, each time, some new *ad hoc* guess for the appropriate $\mu$, whereas optimization is systematic. While optimization here offers only a small improvement, it is an improvement: there is slightly greater precision, with smaller expected errors that shrink more rapidly with increasing $k$. It is also noteworthy that while $a_{\overline{\text{MS}}}$ slightly increases with $k$, the optimized couplant $\bar{a}$ shrinks, consistent with the "induced convergence" scenario.

## Example 1: $Q/\tilde{\Lambda}_{\overline{\text{MS}}} = 340$.

Table 10.1. Results for $\mathcal{R}$, the QCD corrections to $R_{e^+e^-}$, in $(k+1)$th-order ($N^k$LO) at an energy $Q/\tilde{\Lambda}_{\overline{\text{MS}}} = 340$. The upper and lower subtables list, respectively, the $\overline{\text{MS}}$ and optimized results. The columns give the couplant value, the rough form of the series, and the result for $\mathcal{R}$ with an error estimate corresponding to $|r_k a^{k+1}|$, the magnitude of the last term included in the perturbation series.

| Order | $a_{\overline{\text{MS}}}$ | $\mathcal{R}_{\overline{\text{MS}}}$ series | $\mathcal{R}_{\overline{\text{MS}}}$ |
|---|---|---|---|
| $k = 1$ | 0.0381237 | $0.04(1 + 0.05)$ | $0.04017[205]$ |
| $k = 2$ | 0.0382058 | $0.04(1 + 0.05 - 0.02)$ | $0.03955[71]$ |
| $k = 3$ | 0.0382161 | $0.04(1 + 0.054 - 0.019 - 0.004)$ | $0.03939[17]$ |

| Order | $\bar{a}$ | $\bar{\mathcal{R}}$ series | $\bar{\mathcal{R}}$ |
|---|---|---|---|
| $k = 1$ | 0.0414570 | $0.04(1 - 0.02)$ | $0.04043[103]$ |
| $k = 2$ | 0.0394420 | $0.04(1 - 0.01 + 0.01)$ | $0.03944[47]$ |
| $k = 3$ | 0.0391507 | $0.04(1 + 0.003 + 0.002 + 0.001)$ | $0.03941[4]$ |

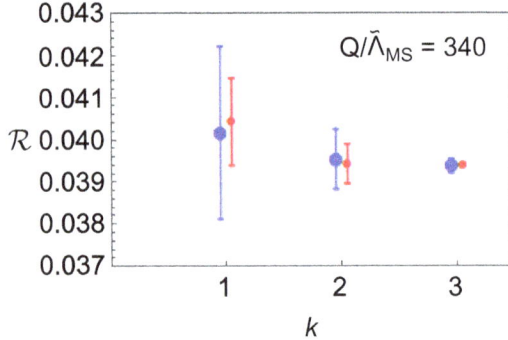

Fig. 10.1. Results for $\mathcal{R}$, the QCD corrections to $R_{e^+e^-}$, in $(k + 1)$th-order ($N^k$LO) at an energy $Q/\tilde{\Lambda}_{\overline{\text{MS}}} = 340$. The larger, blue points displaced leftwards are in the $\overline{\text{MS}}$ scheme, while the smaller, red points displaced rightwards are the optimized results. In both cases, the error bars correspond to $|r_k a^{k+1}|$.

**Example 2:** $Q/\tilde{\Lambda}_{\overline{\text{MS}}} = 68$.

Table 10.2. Results for $Q/\tilde{\Lambda}_{\overline{\text{MS}}} = 68$.

| Order | $a_{\overline{\text{MS}}}$ | $\mathcal{R}_{\overline{\text{MS}}}$ series | $\mathcal{R}_{\overline{\text{MS}}}$ |
|---|---|---|---|
| $k=1$ | 0.0507097 | $0.05(1+0.07)$ | $0.05433[362]$ |
| $k=2$ | 0.0509032 | $0.05(1+0.07-0.03)$ | $0.05287[169]$ |
| $k=3$ | 0.0509356 | $0.05(1+0.07-0.03-0.01)$ | $0.05236[54]$ |
| Order | $\bar{a}$ | $\bar{\mathcal{R}}$ series | $\bar{\mathcal{R}}$ |
| $k=1$ | 0.0568587 | $0.06(1-0.03)$ | $0.05496[190]$ |
| $k=2$ | 0.0525541 | $0.05(1-0.02+0.02)$ | $0.05256[112]$ |
| $k=3$ | 0.0520416 | $0.05(1+0.002+0.003+0.002)$ | $0.05245[13]$ |

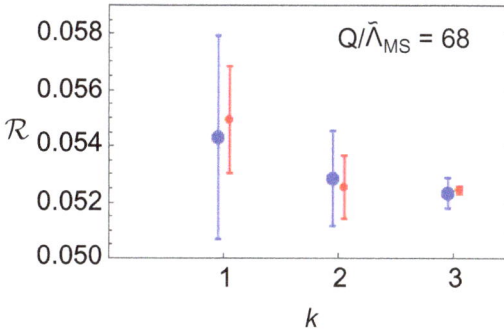

Fig. 10.2. Results for $Q/\tilde{\Lambda}_{\overline{\text{MS}}} = 68$.

## 10.4. Low-Energy Examples

Next we turn to lower-energy examples, where the differences between $\overline{\text{MS}}$ and OPT become more dramatic. With $n_f = 2$ the $\beta$-function's leading coefficients are

$$b = \frac{29}{6}, \qquad c = \frac{115}{58}, \qquad (10.12)$$

and

$$c_2^{\overline{MS}} = \frac{48241}{8352} = 5.77598, \tag{10.13}$$

$$c_3^{\overline{MS}} = \frac{18799309}{902016} + \frac{68881}{12528}\zeta_3 = 27.45054. \tag{10.14}$$

The $\mathcal{R}$ coefficients, in the $\overline{MS}$ scheme, are

$$r_1^{\overline{MS}} = 1.755117, \qquad r_2^{\overline{MS}} = -9.14055, \qquad r_3^{\overline{MS}} = -123.18799. \tag{10.15}$$

(Again, the exact values were used in our calculations.) Inserting these values in Eq. (7.31) yields

$$\rho_2 = -9.92498, \qquad \rho_3 = -115.21021. \tag{10.16}$$

Examples 3–7 give results at successively lower energies; $Q/\tilde{\Lambda}_{\overline{MS}} = 5, 2, 1.7, 1.5$ and $0$. One sees in the $\overline{MS}$ results the characteristic symptoms of an asymptotic series; after initially seeming to converge, the series starts to go bad, with the error estimate *increasing* with order. In Example 3, the effect is just visible in the $k = 3$ result, but it becomes more dramatic in Examples 4 and 5. In Example 6, there is no $k = 3$ $\overline{MS}$ result at all since there is no positive-$a$ solution to the $k = 3$ int-$\beta$ equation. At still lower values of $Q/\tilde{\Lambda}_{\overline{MS}}$, the $k = 2$ and $k = 1$ $\overline{MS}$ int-$\beta$ equations have no acceptable solution.

In contrast, the optimized results show a monotonic decrease in the expected error at higher orders. The $k = 1$ results, in Examples 5 and 6 particularly, are very uncertain at low energies — indeed, for $Q/\tilde{\Lambda}_{\overline{MS}} < 1.438$ (corresponding to $\rho_1(Q) < 0$) there is no solution to the $k = 1$ optimal int-$\beta$ equation, Eq. (8.12). However, for $k = 2$ and $3$ the optimized results improve very significantly — and continue smoothly down to zero energy.

**A Landau pole** is said to occur when the solution for $a$ goes to infinity at some finite $Q$ value and becomes negative at smaller $Q$. In $\overline{MS}$ the Landau pole is at $Q = \tilde{\Lambda}_{\overline{MS}}$ at $k = 1$, and at $Q/\tilde{\Lambda}_{\overline{MS}} = 1.396$ and $1.645$ at $k = 2, 3$, respectively. In OPT, because $c$ here is positive, there is a Landau pole when $k = 1$, but it is absent for $k = 2, 3$. These matters will be discussed in the next chapter.

## Example 3: $Q/\tilde{\Lambda}_{\overline{\text{MS}}} = 5$.

Table 10.3.  Results for $Q/\tilde{\Lambda}_{\overline{\text{MS}}} = 5$.

| Order | $a_{\overline{\text{MS}}}$ | $\mathcal{R}_{\overline{\text{MS}}}$ series | $\mathcal{R}_{\overline{\text{MS}}}$ |
|---|---|---|---|
| $k = 1$ | 0.0862557 | $0.09(1 + 0.15)$ | $0.099[13]$ |
| $k = 2$ | 0.0902494 | $0.09(1 + 0.16 - 0.07)$ | $0.098[7]$ |
| $k = 3$ | 0.0911287 | $0.09(1 + 0.16 - 0.08 - 0.09)$ | $0.090[8]$ |

| Order | $\bar{a}$ | $\bar{\mathcal{R}}$ series | $\bar{\mathcal{R}}$ |
|---|---|---|---|
| $k = 1$ | 0.117285 | $0.12(1 - 0.09)$ | $0.106[11]$ |
| $k = 2$ | 0.0952429 | $0.10(1 - 0.05 + 0.05)$ | $0.095[5]$ |
| $k = 3$ | 0.0899359 | $0.09(1 - 0.01 - 0.02 + 0.04)$ | $0.091[4]$ |

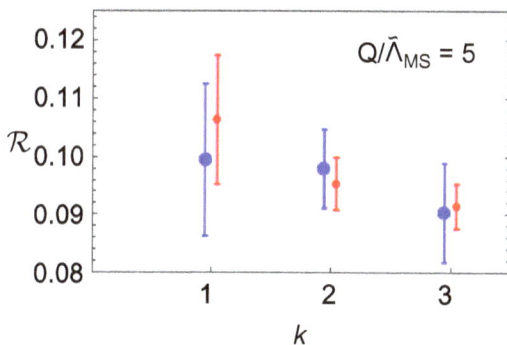

Fig. 10.3.  Results for $Q/\tilde{\Lambda}_{\overline{\text{MS}}} = 5$.

## Example 4: $Q/\tilde{\Lambda}_{\overline{\mathrm{MS}}} = 2$.

Table 10.4.    Results for $Q/\tilde{\Lambda}_{\overline{\mathrm{MS}}} = 2$.

| Order | $a_{\overline{\mathrm{MS}}}$ | $\mathcal{R}_{\overline{\mathrm{MS}}}$ series | $\mathcal{R}_{\overline{\mathrm{MS}}}$ |
|---|---|---|---|
| $k = 1$ | 0.1626471 | $0.16(1 + 0.29)$ | $0.209[46]$ |
| $k = 2$ | 0.1963533 | $0.20(1 + 0.34 - 0.35)$ | $0.195[69]$ |
| $k = 3$ | 0.2193679 | $0.22(1 + 0.39 - 0.44 - 1.30)$ | $-0.08 \pm 0.29$ |
| Order | $\bar{a}$ | $\bar{\mathcal{R}}$ series | $\bar{\mathcal{R}}$ |
| $k = 1$ | 0.3648099 | $0.36(1 - 0.21)$ | $0.288[77]$ |
| $k = 2$ | 0.1725913 | $0.17(1 - 0.17 + 0.18)$ | $0.173[31]$ |
| $k = 3$ | 0.1421756 | $0.14(1 - 0.08 - 0.09 + 0.17)$ | $0.143[25]$ |

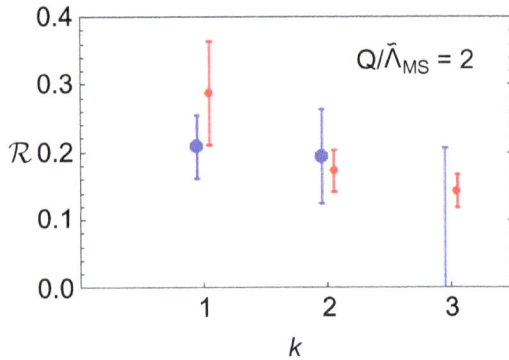

Fig. 10.4.    Results for $Q/\tilde{\Lambda}_{\overline{\mathrm{MS}}} = 2$. (The $k = 3$ $\overline{\mathrm{MS}}$ result is slightly negative; only its error bar is visible.)

## Example 5: $Q/\tilde{\Lambda}_{\overline{MS}} = 1.7$.

Table 10.5.   Results for $Q/\tilde{\Lambda}_{\overline{MS}} = 1.7$.

| Order | $a_{\overline{MS}}$ | $\mathcal{R}_{\overline{MS}}$ series | $\mathcal{R}_{\overline{MS}}$ |
|-------|------|------|------|
| $k = 1$ | 0.1966624 | $0.20(1 + 0.35)$ | $0.265[68]$ |
| $k = 2$ | 0.2691684 | $0.27(1 + 0.47 - 0.66)$ | $0.218[178]$ |
| $k = 3$ | 0.4153849 | $0.42(1 + 0.73 - 1.58 - 8.83)$ | $-3.60 \pm 3.67$ |
| Order | $\bar{a}$ | $\bar{\mathcal{R}}$ series | $\bar{\mathcal{R}}$ |
| $k = 1$ | 0.6669931 | $0.67(1 - 0.28)$ | $0.477[190]$ |
| $k = 2$ | 0.1970393 | $0.20(1 - 0.24 + 0.25)$ | $0.199[49]$ |
| $k = 3$ | 0.1530735 | $0.15(1 - 0.11 - 0.11 + 0.22)$ | $0.153[34]$ |

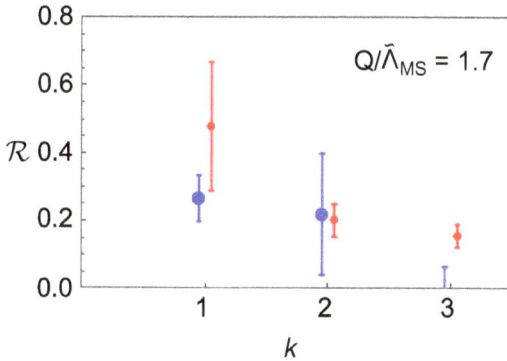

Fig. 10.5.   Results for $Q/\tilde{\Lambda}_{\overline{MS}} = 1.7$. (Only the tip of the huge error bar for the $k = 3$ $\overline{MS}$ result is visible.)

## Example 6: $Q/\tilde{\Lambda}_{\overline{\mathrm{MS}}} = 1.5$.

Table 10.6.   Results for $Q/\tilde{\Lambda}_{\overline{\mathrm{MS}}} = 1.5$.

| Order | $a_{\overline{\mathrm{MS}}}$ | $\mathcal{R}_{\overline{\mathrm{MS}}}$ series | $\mathcal{R}_{\overline{\mathrm{MS}}}$ |
|---|---|---|---|
| $k = 1$ | 0.236877 | $0.24(1 + 0.42)$ | $0.335[98]$ |
| $k = 2$ | 0.431322 | $0.43(1 + 0.76 - 1.70)$ | $0.02 \pm 0.73$ |
| $k = 3$ | no solution | | |

| Order | $\bar{a}$ | $\bar{\mathcal{R}}$ series | $\bar{\mathcal{R}}$ |
|---|---|---|---|
| $k = 1$ | 2.4690661 | $2.5(1 - 0.42)$ | $1.4 \pm 1.0$ |
| $k = 2$ | 0.2173977 | $0.22(1 - 0.31 + 0.33)$ | $0.221[71]$ |
| $k = 3$ | 0.1605183 | $0.16(1 - 0.13 - 0.13 + 0.26)$ | $0.161[42]$ |

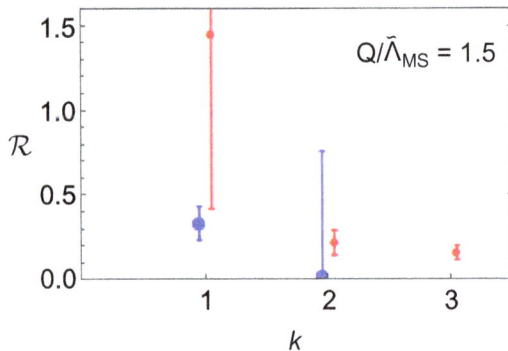

Fig. 10.6.   Results for $Q/\tilde{\Lambda}_{\overline{\mathrm{MS}}} = 1.5$. (There is no $k = 3$ $\overline{\mathrm{MS}}$ result in this case.)

**Example 7:** $Q = 0$ (fixed point).

Table 10.7. Results for the infrared fixed-point limit, $Q = 0$. There are no $\overline{\text{MS}}$ results in this case.

| Order | $\bar{a}$ | $\mathcal{R}$ series | $\bar{\mathcal{R}}$ |
|---|---|---|---|
| $k = 1$ | no solution | — | — |
| $k = 2$ | 0.2635259 | $0.26(1 - 0.76 + 1.01)$ | 0.330[267] |
| $k = 3$ | 0.1800794 | $0.18(1 - 0.25 - 0.16 + 0.44)$ | 0.185[79] |

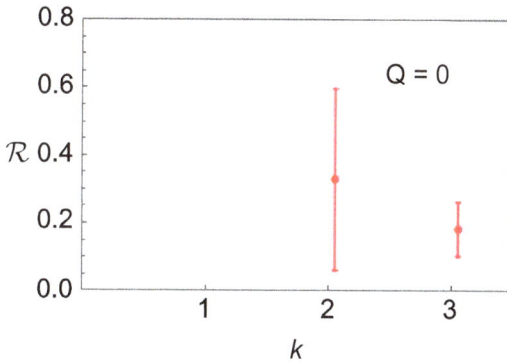

Fig. 10.7. Results for the infrared fixed-point limit, $Q = 0$. There are no $\overline{\text{MS}}$ results in this case, and no $k = 1$ optimized result.

## 10.5. Discussion

Let us summarize the lessons of these numerical examples, which compared the $\overline{\text{MS}}(\mu = Q)$ and optimized results. At moderately high energies, the differences are small — well within the error estimates. The main advantage of optimization here is to achieve better precision — and, very importantly, to have a systematic method that applies to other physical quantities, without the need for new *ad hoc* choices of $\mu$ in each case.

At low energies, however, there are more striking differences. While the optimized results show steady convergence, the $\overline{\text{MS}}$ results begin to exhibit the typical pathologies of a divergent asymptotic series (compare, for example, with Fig. 5.3). One can expect those pathologies to show up in the $\overline{\text{MS}}$ results at higher energies when the series is taken to high enough order. In this sense, the low-energy examples are a "preview" of the divergent-series problems to be expected in $\overline{\text{MS}}$ or any fixed RS.

Whether or not the optimized results in QCD will exhibit "induced convergence" is a matter of conjecture at present. The examples here certainly show a consistent shrinking of the optimized couplant from one order to the next.

The good convergence of the optimized results remains true even in the $Q \to 0$ limit, where the third-order finding of a limit $\mathcal{R} \to 0.3 \pm 0.3$ is confirmed and made more precise; $\mathcal{R} \to 0.2 \pm 0.1$. This result is important because there are many indications from phenomenology that the QCD couplant does "freeze" at low energies. Usually freezing is something put in by hand, but here it is an outcome, a *prediction*. There is nothing in the optimization approach that forces freezing to occur; the fact that it does for $R_{e^+e^-}$ is due to the $\rho_2, \rho_3$ values resulting from the Feynman-diagram calculations. The next chapter will discuss the topic of the infrared limit in more detail and more generality.

## Appendix 10.A: $\beta$ Function and $\mathcal{R}_{e^+e^-}$ Coefficients in $\overline{\text{MS}}$

The $\beta$-function coefficients, $bc_j$, and the $\mathcal{R}_{e^+e^-}$ coefficients, $r_i$, in the $\overline{\text{MS}}$ scheme are polynomials in $n_f$, with each fermion loop in a diagram being associated with an $n_f$ factor. We choose to swap $n_f$ for $b$

$$b = \frac{33 - 2n_f}{6}, \tag{10A.1}$$

both because it makes the results a little more compact and because it is convenient when investigating the large-$b$ and small-$b$ (Banks-Zaks) approximations. Here $\zeta_s$ is the Riemann zeta-function. References to the original calculations are given in the Bibliography.

The $\beta$-function coefficients, in the $\overline{\text{MS}}$ prescription, are as follows:

$$c = -\left(\frac{107}{8}\right)\frac{1}{b} + \frac{19}{4},$$

$$c_2^{\overline{\text{MS}}} = \left(-\frac{37117}{768}\right)\frac{1}{b} + \frac{243}{32} + \frac{325}{192}b,$$

$$c_3^{\overline{\text{MS}}} = \left(\frac{53981}{1152} + \frac{5335}{32}\zeta_3\right)\frac{1}{b} + \left(-\frac{1544327}{13824} - \frac{16171}{288}\zeta_3\right)$$

$$+ \left(\frac{2587}{96} + \frac{809}{144}\zeta_3\right)b - \frac{1093}{3456}b^2.$$

The $\mathcal{R}_{e^+e^-}$ series coefficients, in the $\overline{\text{MS}}(\mu = Q)$ scheme, are as follows:

$$r_1^{\overline{\text{MS}}} = \frac{1}{12} + \left(\frac{11}{4} - 2\zeta_3\right)b,$$

$$r_2^{\overline{\text{MS}}} = \left(-\frac{12521}{288} + 13\zeta_3\right) + \left(\frac{401}{24} - \frac{53}{3}\zeta_3 + \frac{25}{3}\zeta_5\right)b$$

$$+ \left(\frac{151}{18} - \frac{19}{3}\zeta_3 - \frac{1}{2}\zeta_2\right)b^2 + \left(\frac{(\sum q_i)^2}{3\sum q_i^2}\right)\left(\frac{55}{72} - \frac{5}{3}\zeta_3\right),$$

$$r_3^{\overline{\text{MS}}} = \left(-\frac{3963761}{20736} + \frac{677833}{3456}\zeta_3 - \frac{275}{24}\zeta_5\right)$$

$$+ \left(-\frac{38969}{128} + \frac{535}{32}\zeta_2 + \frac{6907}{96}\zeta_3 + \frac{165}{2}\zeta_3^2 + \frac{9595}{144}\zeta_5 - \frac{665}{24}\zeta_7\right)b$$

$$+ \left(\frac{236089}{1728} - \frac{97}{16}\zeta_2 - \frac{13859}{96}\zeta_3 + \frac{15}{2}\zeta_3^2 + \frac{445}{12}\zeta_5\right)b^2$$

$$+ \left(\frac{6131}{216} - \frac{33}{8}\zeta_2 - \frac{203}{12}\zeta_3 + 3\zeta_2\zeta_3 - \frac{15}{2}\zeta_5\right)b^3$$

$$+ \left(\frac{(\sum q_i)^2}{3\sum q_i^2}\right)\left(\frac{995}{576} - \frac{905}{72}\zeta_3 + \frac{175}{36}\zeta_5\right.$$

$$+ \left.\left(\frac{745}{144} - \frac{65}{8}\zeta_3 - \frac{5}{2}\zeta_3^2 + \frac{25}{4}\zeta_5\right)b\right).$$

The $\rho_2, \rho_3$ invariants for $\mathcal{R}_{e^+e^-}$ are therefore

$$\rho_2 = \left(-\frac{12087}{256}\right)\frac{1}{b} + \left(\frac{143}{288} - \frac{55}{4}\zeta_3\right)$$

$$+ \left(\frac{937}{192} - \frac{47}{6}\zeta_3 + \frac{25}{3}\zeta_5\right)b + \left(\frac{119}{144} - \frac{1}{2}\zeta_2 + \frac{14}{3}\zeta_3 - 4\zeta_3^2\right)b^2$$

$$+ \left(\frac{(\sum q_i)^2}{3\sum q_i^2}\right)\left(\frac{55}{72} - \frac{5}{3}\zeta_3\right),$$

$$\rho_3 = \left(\frac{252613}{4608} + \frac{5335}{32}\zeta_3\right)\frac{1}{b} + \left(-\frac{2217005}{10368} + \frac{121613}{864}\zeta_3 - \frac{275}{12}\zeta_5\right)$$

$$- \left(\frac{7913}{576} - \frac{535}{16}\zeta_2 - \frac{3619}{9}\zeta_3 - \frac{535}{2}\zeta_3^2 - \frac{9295}{72}\zeta_5 - \frac{665}{12}\zeta_7\right)b$$

$$+ \left(\frac{15683}{576} - \frac{95}{8}\zeta_2 + \frac{3599}{24}\zeta_3 - 174\zeta_3^2 - \frac{190}{3}\zeta_5 + 100\zeta_3\zeta_5\right)b^2$$

$$+ \left(\frac{665}{432} - \frac{61}{6}\zeta_3 + 56\zeta_3^2 - 32\zeta_3^3 - 15\zeta_5\right)b^3$$

$$+ \left(\frac{(\sum q_i)^2}{3\sum q_i^2}\right)\left(\frac{295}{96} - \frac{875}{36}\zeta_3 + \frac{175}{18}\zeta_5\right)$$

$$+ \left(-\frac{325}{144} + \frac{245}{12}\zeta_3 - 25\zeta_3^2 + \frac{25}{2}\zeta_5\right)b\right).$$

For numerical values, neglecting the $\sum q_i$ terms, see Table 11.1 in Chapter 11.

> **Note that**, while $\rho_2, \rho_3, \dots$ and $c$ have definite decompositions in terms of $b$, and hence in terms of $n_f$, the same is *not* true of $\rho_1(Q)$, which involves $\tilde{\Lambda}$, whose $n_f$ dependence could be anything, especially since its specific definition is just a convention.

One more coefficient of the $\beta$ function has been calculated. We cannot make use of it for optimization until the $r_4$ coefficient is calculated:

$$c_4^{\overline{MS}} = \left( \frac{1081830511}{663552} + \frac{17251949}{13824} \zeta_3 - \frac{191675}{192} \zeta_5 \right) \frac{1}{b}$$

$$+ \left( -\frac{1452057293}{1327104} - \frac{48015}{512} \zeta_4 - \frac{4489165}{27648} \zeta_3 + \frac{856625}{2304} \zeta_5 \right)$$

$$+ \left( \frac{33737869}{221184} + \frac{16171}{512} \zeta_4 - \frac{176837}{2304} \zeta_3 - \frac{88415}{2304} \zeta_5 \right) b$$

$$+ \left( \frac{471499}{110592} - \frac{809}{256} \zeta_4 + \frac{39409}{2304} \zeta_3 - \frac{345}{128} \zeta_5 \right) b^2$$

$$+ \left( \frac{1205}{18432} - \frac{19}{64} \zeta_3 \right) b^3.$$

# Part III

# Special Topics

# Chapter 11

# Infrared Limit: Fixed and Unfixed Points

## 11.1. QCD Perturbation Theory in the Infrared?

The ultraviolet region, for an asymptotically-free theory, is the natural domain of perturbation theory: The effective couplant tends to zero as $Q \to \infty$, so when $Q$ is large enough one can confidently expect perturbation theory to give a good approximation. As $Q$ decreases the effective couplant increases and perturbation theory becomes less accurate and less reliable. Moreover, so-called "higher-twist" terms — terms exponentially small in the couplant, and hence suppressed by positive powers of $\tilde{\Lambda}/Q$ — become ever more important. Thus, it may be foolish to consider perturbation theory in the infrared.

On the other hand, it may not be. While the effective couplant grows as $Q$ decreases, it does not necessarily grow without limit, and, in some circumstances, it may remain quite small even in the $Q \to 0$ limit. That, indeed, seems to be the case for $R_{e^+e^-}$ in QCD, as was seen in the numerical examples in the last chapter.

Of course, one should recognize that the infrared behaviour of perturbation theory will not predict the *actual* infrared behaviour of the theory, because of the "higher-twist" terms invisible to perturbation theory. Experimentally, physical quantities, such as $R_{e^+e^-}$, show a lot of structure at low energies, related to hadronic resonances, whereas the perturbative prediction varies smoothly with $Q$ (at least, if quark masses are neglected). Nevertheless, it has long been thought that the perturbative result represents a sort of average through the resonant peaks, so that it can be meaningfully compared with a smoothed or "smeared" version of the experimental data. Such ideas, and the successes of various phenomenological models that invoke a "freezing" of the QCD coupling constant, suggest that low-energy perturbation theory may have some limited predictive power. This chapter will not discuss the phenomenological issues but will study the infrared limit of optimized perturbation theory, motivated by the above considerations.

> **New methods** that introduce variational mass parameters for the quark and/or gluon fields, analogous to the CK expansion, and then optimize these, as well as the RS, offer a very promising approach to the infrared region. While it is beyond the scope of this book to discuss such methods (see references in the bibliography), the lessons learned here in "pure" perturbation theory should be useful background for those endeavours.

One possibility is that there may be no infrared limit at all: There may be a "Landau pole" at some finite $Q$ of order $\tilde{\Lambda}$ where the effective couplant, and hence the perturbative result for $\mathcal{R}$, goes to infinity. Such is the case at first order (a qualitative approximation only), where the effective couplant is $a \approx 1/(b\ln(Q/\tilde{\Lambda}))$. In second order, if the coefficient $c$ is positive (as it is for 8 or fewer quark flavours in QCD), then a Landau pole also occurs. However, if $c$ is negative, then the second-order $\beta$ function, $-ba^2(1+ca)$ vanishes at a positive value $a = -1/c$, and the couplant approaches this value as $Q \to 0$. At higher orders, whatever the sign of $c$, it might happen that the $\beta$ function (in the optimized scheme) has a zero at some value $a = a^*$, and if $a^*$ is quite small one can hope that perturbation theory in the infrared is meaningful.

A simple zero of the $\beta$ function is known as a "fixed point." There is a body of conventional lore about fixed points that we examine next, before turning to an analysis of OPT in the infrared limit.

## 11.2. "Fixed-Point Lore" and Its Limitations

The key properties of a renormalizable field theory, according to many accounts, follow simply from a graph of its $\beta$ function.

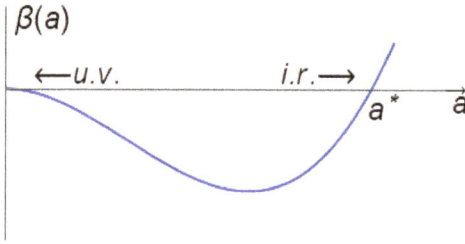

Fig. 11.1.   Conventional sketch of "the $\beta$ function" in an asymptotically free theory with an infrared fixed point. The couplant flows to zero in the ultraviolet and to $a^*$ in the infrared.

Figure 11.1, for instance, supposedly represents an asymptotically free theory with an infrared fixed point at $a = a^*$.

The problem with this lore is that there is no such thing as *the $\beta$ function*! The $\beta$ function is not a physical quantity; it is not a unique, well-defined function characterizing the theory. It is RP dependent, and — unlike the case of physical quantities — its RP dependence would remain an issue even if we could magically calculate all orders of perturbation theory.

> **In the literature** there are sometimes claims to have calculated "the exact $\beta$ function" of some theory. However, since $\beta$ is RP dependent, one is always free to define $\beta(a)$ to be any desired function, provided that its power-series expansion starts $-ba^2(1 + ca + O(a^2))$. For example, there is an RP, once used by 't Hooft, in which $\beta(a)$ is *defined* to be $-ba^2(1 + ca)$. Another example of an "exact $\beta$ function" might be $-ba^2/(1 - ca)$, and there are infinitely many other possibilities. Of course, if one could show that some particular RP, associated with some particular form of the all-orders $\beta$ function, also led to some specific all-orders form for one or more physical quantities, then that would be a major achievement. However, without that, claims for an "exact $\beta$ function" are empty.

The RP dependence of the $\beta$ function is, of course, acknowledged in conventional accounts, and the effects of RP transformations

$$a' = a(1 + v_1 a + v_2 a^2 + \cdots),$$
(11.1)

with $\beta$ transforming as

$$\beta'(a') = \frac{da'}{da}\beta(a), \qquad (11.2)$$

are discussed. It is observed that if $\beta(a)$ vanishes at $a = a^*$, then $\beta'(a')$ will vanish at some corresponding $a' = a'^*$ — pictorially, consider a graph of $\beta(a)$ and imagine a stretching, or other distortion, of the horizontal axis. Thus — it would seem — the statement that a certain theory "has a fixed point" is basically an RP-independent one. (Another claim, that the slope of the $\beta$ function at a fixed point is RP invariant, will be discussed later.)

Following those remarks about RP dependence, though, most accounts then proceed as if "the $\beta$ function" were a well-defined, unique function characteristic of the theory — implicitly assuming that one can choose a "good" RP once-and-for-all. However, as stressed in Chapters 4 and 5, with non-invariant approximations it is a mistake to assume that the extraneous variables must have fixed values. Just as the renormalization scale, $\mu$, should "run" with $Q$, the other aspects of the RS can and should be chosen differently in different cases. Because $\beta(a)$ changes under optimization, the usual fixed-point lore, though undoubtedly valuable, will need re-examination.

The fact that we are doing perturbation theory and using *truncations* of the $\beta$ function is crucial. Under a scheme transformation (11.1) the new approximate $\beta$ function is not given by Eq. (11.2), but by a *truncation* thereof. That fact spoils the argument that, generically, $\beta'(a')$ will have a zero if $\beta(a)$ does. Thus, *at finite orders*, fixed points can appear or disappear under RP transformations. The existence or non-existence of a fixed point, being RP dependent in that sense, is not necessarily a property of the theory but may depend on which physical quantity $\mathcal{R}$ is being calculated, and to what order.

**In QCD** with $n_f = 2$ the $\overline{\text{MS}}$ $\beta$ function is $-ba^2(1 + 1.98a + 5.78a^2 + 27.5a^3 + \cdots)$, and has no zero in second, third, or fourth orders. Thus, the $\overline{\text{MS}}$ results must go to infinity at some finite $Q$. However, the optimized $\beta$ function for $\mathcal{R}_{e^+e^-}$ in the $Q \to 0$ limit turns out to be $-ba^2(1 + 1.98a - 22a^2)$ at third order, and $-ba^2(1 + 1.98a - 5.62a^2 - 199a^3)$ at fourth order, giving fixed-point values $a^* = 0.264$ and $0.180$, respectively. Thus, as was seen in the numerical examples of the last chapter, the optimized results for $\mathcal{R}_{e^+e^-}$ continue smoothly down to $Q = 0$. However, there is no guarantee that other physical quantities will necessarily show similar behaviour.

It should not be assumed that this point will become less and less important as we go to higher orders of perturbation theory — because of the divergence of the fixed-scheme power series for $\mathcal{R}$ and $\beta(a)$. While the fixed-scheme $\mathcal{R}$ results will presumably show asymptotic-series behaviour, "settling down" temporarily to a good approximation to the true $\mathcal{R}$, the same need not be true of $\beta(a)$, which is not a physical quantity and has no "true value." In OPT, while we can hope that successive approximations to $\mathcal{R}$ will show "induced convergence," we should not necessarily expect convergence of the associated results for the optimized $\beta$-function: It need not have an infinite-order limit (not even in a re-summed sense) since it is not a physical quantity, and inherently depends on the order index $k$.

The optimized $\beta$ function also inherently depends on what physical quantity $\mathcal{R}$ one is calculating. Thus, it is not necessarily true that the physical quantities in a certain theory either all have, or all do not have, a finite infrared limit. One should be cautious of speaking of the *theory* having, or not having, a fixed point.

One should also re-examine the usual lore that "if a theory has an infrared fixed point then all physical quantities show the same power-law behaviour as $Q \to 0$." The conventional argument goes as follows: Near the fixed point, $\beta(a)$ is approximately linear:

$$\beta(a) \propto a - a^*, \tag{11.3}$$

with the constant of proportionality being the slope of the $\beta$ function at the fixed point:

$$\dot{\beta}^* \equiv \frac{d\beta}{da}\bigg|_{a=a^*}. \tag{11.4}$$

For reasons explained below, $\dot{\beta}^*$ is claimed to be RS invariant. Thus, any RS may be used, and in the EC scheme (assuming for simplicity that $\mathcal{R}$ has $\mathrm{P} = 1$) we have, at low $Q$,

$$Q\frac{d\mathcal{R}}{dQ} = \beta_{\mathrm{EC}}(\mathcal{R}) \approx \dot{\beta}^*(\mathcal{R} - \mathcal{R}^*). \tag{11.5}$$

When integrated, this equation implies a power law

$$\mathcal{R}^* - \mathcal{R} \propto Q^{\gamma^*}, \tag{11.6}$$

with the critical exponent $\gamma^* = \dot{\beta}^*$ being the same for all physical quantities $\mathcal{R}$. (The generalization to any $\mathrm{P}$ is straightforward; see Exercise 11.1.)

The argument that $\dot{\beta}^*$ is RS invariant goes as follows. Consider two RS's, primed and unprimed, whose couplants are related by a general scheme transformation, Eq. (11.1). Formally, their $\beta$ functions are related by Eq. (11.2) and hence the derivative of the $\beta$ function will transform as

$$\frac{d\beta'}{da'} = \frac{d\beta}{da} + \beta(a)\frac{d^2a'}{da^2} \Big/ \frac{da'}{da}. \tag{11.7}$$

Since $\beta(a)$ vanishes at the fixed point, it would seem that $\dot{\beta}^*$ is the same in both primed and unprimed schemes. Careful authors qualify this result with the proviso that $da'/da$ must not vanish and $d^2a'/da^2$ must not be singular at $a = a^*$. Those conditions might seem pedantic, but they cannot be entirely disregarded, as Chýla was the first to warn. (See Exercise 11.2.) The subtleties at the formal level are discussed in Appendix 11.A.

Because of these subtleties we suggest that, within a given theory, there may be different classes of physical quantities: One class might have a Landau-pole, precluding any continuation to $Q = 0$ without explicitly including higher-twist, non-perturbative terms. Another class might have a finite infrared limit, with a common power-law approach to the $Q \to 0$ limit. Other classes might also have finite infrared behaviour, but with a qualitatively different approach to the $Q \to 0$ limit and/or a markedly different infrared couplant.

Furthermore, formal results may or may not apply when some particular approximation method is used. In OPT at any order (except second) it turns out — for non-trivial reasons explained in Appendix 11.B — that the slope of the optimized $\beta$ function at a fixed point does indeed give the $\gamma^*$ exponent governing the $Q \to 0$ behaviour of the optimized approximation to $\mathcal{R}(Q)$. However, it is an open question whether the OPT results for the $\gamma^*$'s of all physical quantities are approximately the same. Moreover, as discussed later on, in OPT a finite infrared limit can occur at an "unfixed point" where $\beta(a)$ does not vanish (but $\partial \mathcal{R}/\partial a$ does). In that case $\mathcal{R}^* - \mathcal{R}$ is not given by a power law, but is proportional to $1/(\ln Q)^2$.

## 11.3. Infrared Behaviour in OPT at Second Order

Before embarking upon a more general analysis of OPT in the infrared, we first examine the case of second order ($k = 1$). Here the $\beta$ function is just $-ba^2(1 + ca)$. When the invariant coefficient $c$ is positive, $\beta(a)$ has no zero: Thus, a Landau pole will occur, with $a$ and the $\mathcal{R}$ result going to infinity at some finite $Q$. However, when $c$ is negative (which would happen in QCD if $n_f > 8\frac{1}{19}$) the $\beta$ function has a zero at $a^* = -1/c$. One might therefore expect typical fixed-point behaviour, with a finite infrared limit $\mathcal{R}^*$ of order $a^{*\mathrm{P}}$. However, in OPT things are more subtle because the optimized $r_1$ coefficient, obtained in Eq. (8.10), is

$$\bar{r}_1 = -\frac{\mathrm{P}}{(\mathrm{P}+1)} \frac{c}{(1+c\bar{a})}, \qquad (11.8)$$

which goes to infinity at the fixed point. Thus, the OPT result for $\bar{\mathcal{R}}^{(2)} = \bar{a}^{\mathrm{P}}(1 + \bar{r}_1\bar{a})$, though finite at any finite $Q$, tends to infinity as $Q \to 0$. At low values of $Q$, of order $\tilde{\Lambda}_{\mathcal{R}}$, where $\rho_1(Q) \approx 0$ there is a plateau, with $\bar{\mathcal{R}}^{(2)}$ of order $a^{*\mathrm{P}}$. It is only at ultra-low $Q$, when $\rho_1(Q)$ becomes large and negative, that there is a narrow "spike" proportional to $-\ln Q$ as $Q \to 0$. This behaviour is illustrated in Fig. 11.2.

Of course, the error estimate also diverges as $Q \to 0$, so the height and width of the spike are extremely uncertain. Similar "spiking," for $n_f \gtrsim 8$, occurs in higher orders, as we shall see, though there the spike does not extend to infinity.

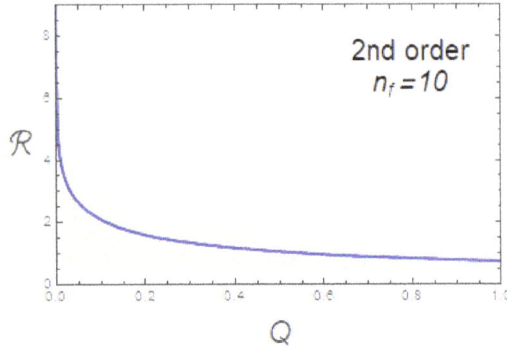

Fig. 11.2.   Second-order OPT results for $\mathcal{R}_{e^+e^-}$ with $n_f = 10$ at low energies. The energy $Q$ is in units of $\tilde{\Lambda}_{\mathcal{R}}$. The shaded region indicates the error estimate. In this case $\mathrm{P} = 1$ and $a^* = -1/c = 26/37 \approx 0.7$. There is a plateau at $Q \sim \tilde{\Lambda}_{\mathcal{R}}$ where $\mathcal{R}$ is roughly $a^*$, but there is a spike, extending to infinity, as $Q \to 0$.

## 11.4.  Finite Infrared Limits in OPT: Two Mechanisms

Beyond second order, the optimization procedure involves determining optimal values for the higher-order $\beta$-function coefficients, $c_2, c_3, \ldots$, and these evolve as the energy $Q$ is changed. Thus, the optimized $\beta$ function is not a fixed function of $a$, but itself evolves with $Q$.

In some cases, this evolution makes little qualitative difference and the infrared limit of OPT arises in basically the usual way from a simple zero of the (optimized) $\beta$ function. In QCD with $n_f = 2$ one finds such behaviour for $\mathcal{R}_{e^+e^-}$ in both third and fourth orders, as seen in the last numerical example of the previous chapter. The fourth-order result, as a function of $Q$ is shown in Fig. 11.3. The approach to the $Q \to 0$ limit is characterized by a $\gamma^*$ exponent of about 3, so that $\mathcal{R}$ is nearly constant at low energies.

> **When the $\gamma^*$ exponent** is greater than 1 it is natural to speak of a "freezing" of the effective couplant. When $\gamma^*$ is less than one, however, it would be more descriptive to speak of a "spiking" of the couplant in the infrared limit.

In some cases, however, the evolution of the optimized $\beta$ function with $Q$ plays a crucial role and a finite infrared limit in OPT occurs in a quite different way. This was quite unanticipated, and turned up

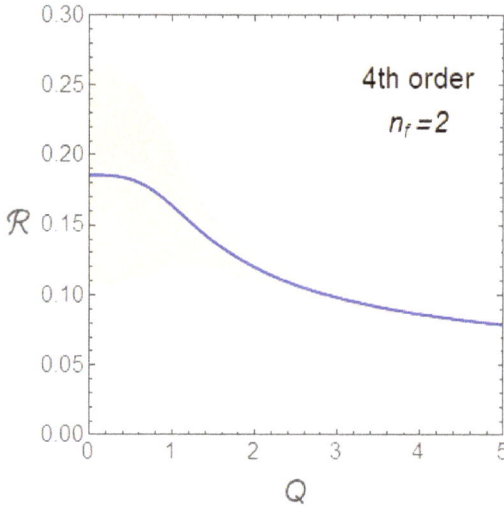

Fig. 11.3. Fourth-order OPT results for $\mathcal{R}_{e^+e^-}$ for $n_f = 2$. The energy $Q$ is in units of $\tilde{\Lambda}_{\mathcal{R}}$. The shaded region indicates the error estimate. This case illustrates infrared "freezing" of the couplant, due to a fixed point of the optimized $\beta$ function.

in numerical investigations of QCD at higher $n_f$ values. The infrared limit here arises by a "pinch mechanism." The evolving $\beta$ function develops a minimum that, as $Q \to 0$, just touches the axis at $a_p$ (the "pinch point"), while the infrared limit of the optimized couplant is at a larger value, $a^*$ (the "unfixed point"). This mechanism produces an "extreme" or "logarithmic" spiking of the couplant as $Q \to 0$; see Fig. 11.4.

The next sections examine the fixed-point mechanism and pinch mechanism, respectively.

## 11.5. Fixed-Point Mechanism

A finite $Q \to 0$ limit for $\mathcal{R}(Q)$ can occur by essentially the familiar fixed-point mechanism, with the optimized $B(a)$ function manifesting a simple zero at $a = a^*$ (see Fig. 11.5). The limiting behaviour can be analyzed as follows. For $a$ close to $a^*$ one can linearize $B(a)$ as

$$B(a) \approx \sigma(a^* - a), \tag{11.9}$$

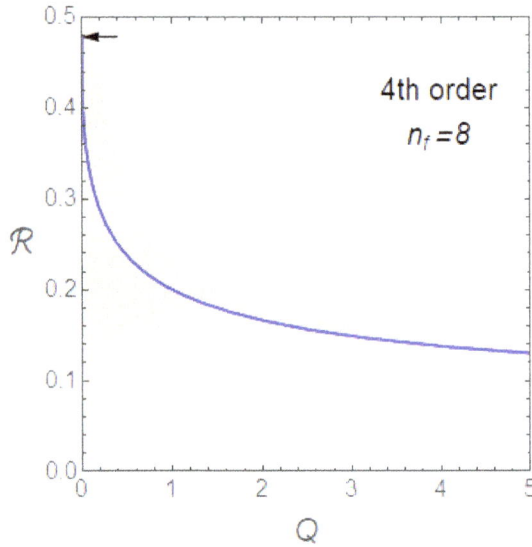

Fig. 11.4.   As Fig. 11.3 but for $n_f = 8$. The arrow indicates the infrared limit. This case illustrates "logarithmic spiking" of the couplant due to the "pinch" mechanism discussed in Sec. 11.6.

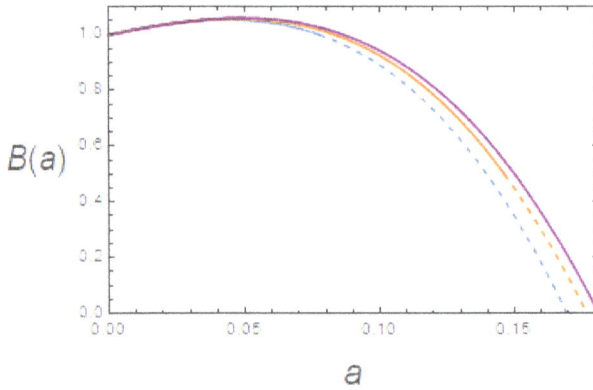

Fig. 11.5.   The evolving optimized $B(a) \equiv \beta(a)/(-ba^2)$ function at fourth order for $n_f = 2$. The upper, solid curve is the $Q = 0$ limiting form with a fixed point at $a^* = 0.180844$. The two lower curves correspond to larger $Q$ values, and are shown dashed when $a > \bar{a}$.

where $\sigma$ is some positive constant (directly related to the slope of the $\beta$ function at its fixed point; $\dot{\beta}^* = ba^{*2}\sigma$). The integrals $I_j(a)$ of Eq. (7.10) will then diverge in the infrared limit, $a \to a^*$:

$$I_j(a) \to \int_0^a dx \frac{x^{j-2}}{\sigma^2(a^* - x)^2} \to \frac{a^{*j-2}}{\sigma^2} \frac{1}{(a^* - a)}. \tag{11.10}$$

Substituting in $B_j(a)$, Eq. (7.9), one finds that the $\frac{1}{(a^*-a)}$ factor is cancelled by the $(a^* - a)$ factor in $B(a)$, yielding

$$B_j(a) \to \frac{(j-1)}{\sigma a^*}. \tag{11.11}$$

This result corresponds to

$$\left. \frac{\partial a}{\partial c_j} \right|^* = \frac{a^{*j}}{\sigma}, \tag{11.12}$$

which indeed follows directly by asking how the root $a^*$ of the equation $\sum_i c_i^* a^{*i} = 0$ changes as one specific $c_j^*$ is varied (see Exercise 11.3). The slope parameter $\sigma$ is given by

$$\sigma = -B'(a)\big|_{a=a^*} = -\sum_{j=0}^{k} jc_j^* a^{*j-1} = \sum_{j=0}^{k-1} (k-j)c_j^* a^{*j-1}, \tag{11.13}$$

where the last step uses the fixed-point condition $\sum_{j=0}^{k} c_j^* a^{*j} = 0$ to eliminate $c_k^*$.

With the limiting $B_j$'s from Eq. (11.11) one can construct the limiting $H_i$'s:

$$H_i \to \frac{1}{\sigma a^*} \sum_{j=0}^{k-i} c_j^* a^{*j} (i - j - 1). \tag{11.14}$$

Hence, in the fixed-point limit, the formula for the $s_m \equiv \left(\frac{P+m}{P}\right) r_m$ coefficients, Eq. (9.13), becomes (for $k > 1$)

$$s_m^* a^{*m} = \frac{1}{(k-1)} \left( \sum_{j=0}^{m} c_j^* a^{*j} (k - m - j - 1) - \sum_{j=0}^{m-1} c_j^* a^{*j} (k - m - j) \right). \tag{11.15}$$

Writing $k - m - j$ as $k - 2m + (m - j)$ and noting that the $(m - j)$ terms will cancel between the two summations, one may simplify the result to

$$\hat{s}_m = \frac{1}{(k-1)} \left[ (k - 2m - 1)\hat{t}_m - (k - 2m)\hat{t}_{m-1} \right], \tag{11.16}$$

or, equivalently,

$$\hat{s}_m = \frac{1}{(k-1)} \left[ (k - 2m)\hat{c}_m - \hat{t}_m \right], \tag{11.17}$$

where

$$\hat{s}_m \equiv s_m^* a^{*m}, \qquad \hat{c}_m \equiv c_m^* a^{*m}, \qquad \hat{t}_m \equiv \sum_{i=0}^{m} \hat{c}_j. \tag{11.18}$$

One may substitute Eq. (11.17) for the optimal-scheme $r_m^*$'s into the expressions for the $\rho_i$ invariants. From the resulting $\rho_2$ one can solve for the optimal-scheme $c_2^*$ in terms of $a^*, c, \rho_2$. Then, making use of that result, one may solve for $c_3^*$ in terms of $a^*, c, \rho_2, \rho_3$, and so on up to $c_k^*$. Substituting in the fixed-point condition $B(a^*) = 0$ then produces an equation for $a^*$ that involves only the invariants $c, \rho_2, \ldots, \rho_k$. One can then find $a^*$ numerically as the smallest positive root of that equation. Finally, the expressions for the $c_j^*$'s in terms of $a^*$ and the invariants can be substituted back into Eq. (11.17) to determine the $r_m^*$'s. Hence, one can find $\mathcal{R}^*$ in terms of invariant quantities.

We now specialize to low orders and to the $P = 1$ case.

**At third order** $(k = 2)$ one finds from Eq. (11.17)

$$\hat{s}_1 = -(1 + \hat{c}), \qquad \hat{s}_2 = -2\hat{c}_2, \tag{11.19}$$

and hence

$$r_1^* = -\frac{(1 + ca^*)}{2a^*}, \qquad r_2^* = -\frac{2}{3} c_2^*. \tag{11.20}$$

Substituting in the definition of $\rho_2$ and solving for $c_2^*$ yields

$$c_2^* = \frac{3}{4a^{*2}} (1 + (4\rho_2 - c^2)a^{*2}). \tag{11.21}$$

Then substituting in the fixed-point condition $B(a^*) = 1 + ca^* + c_2^* a^{*2} = 0$ gives

$$\frac{7}{4} + ca^* + 3\left(\rho_2 - \frac{1}{4}c^2\right)a^{*2} = 0. \tag{11.22}$$

The relevant $a^*$ is the smallest positive root of this equation. The result for $\mathcal{R}^*$ at third order can then be expressed as

$$\mathcal{R}^* = a^*\left(\frac{7}{6} + \frac{1}{6}ca^*\right). \tag{11.23}$$

**At fourth order** $(k = 3)$ one obtains

$$\hat{s}_1 = -\frac{1}{2}, \quad \hat{s}_2 = -\frac{1}{2}(1 + \hat{c} + 2\hat{c}_2), \quad \hat{s}_3 = -\frac{3}{2}\hat{c}_3, \tag{11.24}$$

and hence

$$r_1^* = -\frac{1}{4a^*}, \quad r_2^* = -\frac{(1 + ca^* + 2c_2^* a^{*2})}{6a^{*2}}, \quad r_3^* = -\frac{3}{8}c_3^*. \tag{11.25}$$

By substituting in the definitions of $\rho_2, \rho_3$, one can then find $c_2^*, c_3^*$ in terms of $a^*$ and those invariants:

$$c_2^* = \frac{11 - 4ca^* + 48\rho_2 a^{*2}}{32a^{*2}}, \tag{11.26}$$

$$c_3^* = \frac{5 + 3ca^* + 16\rho_3 a^{*3}}{4a^{*3}}. \tag{11.27}$$

The fixed-point condition can then be expressed entirely in terms of invariants as

$$\frac{83}{32} + \frac{13}{8}ca^* + \frac{3}{2}\rho_2 a^{*2} + 4\rho_3 a^{*3} = 0. \tag{11.28}$$

The final result for the limiting value of $\mathcal{R}$ at fourth order can then be simplified to

$$\mathcal{R}^* = a^*\left(\frac{249}{256} + \frac{13}{64}ca^* + \frac{1}{16}\rho_2 a^{*2}\right). \tag{11.29}$$

Results for the $\mathcal{R}_{e+e^-}$ case will be discussed in Sec. 11.7. At third order fixed-point behaviour is found at all $n_f$'s, though for low $n_f$ the error estimates are large. At fourth order a fixed point is found for

$n_f$ up to 6, and also for $n_f = 16$. However, there is no positive root of Eq. (11.28) when $n_f$ is $7, \ldots, 14$, while for $n_f = 15$ a positive root exists but it gives a negative $\sigma$, which is unacceptable (see Exercise 11.10(iii)). Nevertheless, at those $n_f$'s the optimization procedure yields results that remain bounded as $Q \to 0$; it does so, however, by a quite different mechanism.

## 11.6. Pinch Mechanism

The essence of the pinch mechanism is illustrated in Fig. 11.6, which shows the evolution of the optimized $B(a)$ function in the $n_f = 8$ case. As $Q$ is lowered the optimized $c_2, c_3$ coefficients change so that $B(a)$ develops a minimum — which, in the limit $Q \to 0$, just touches the horizontal axis at a "pinch point," $a_\mathrm{p}$. Although this point is then a double root of $B(a) = 0$, it does *not* represent a fixed point. The infrared-limit of the optimized couplant is not $a_\mathrm{p}$ but a larger value, $a^\star$, dubbed the "unfixed point" to stress that it is not a zero of the $\beta$ function.

> **The infrared limits** of $a$ or the $r_m$, $c_j$ coefficients will be indicated by $\star$ rather than $*$ when they arise from an unfixed point, not a fixed point.

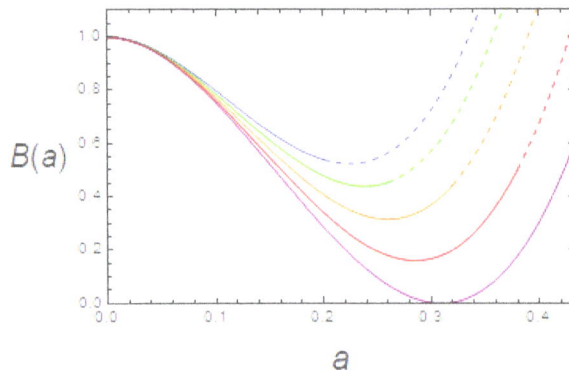

Fig. 11.6. The evolving optimized $B(a) \equiv \beta(a)/(-ba^2)$ function at fourth order for $n_f = 8$. The curves, from top to bottom, are for descending $Q$ values. They are shown dashed for $a > \bar{a}$. The lowest curve is the infrared-limiting form, with the pinch point at $a_\mathrm{p} = 0.3094$ and the unfixed point at $a^\star = 0.432267$.

One can understand this infrared behaviour analytically as follows. $B(a)$ can be approximated around its minimum (at, or nearly at, the pinch point $a_\mathrm{p}$) by

$$B(a) \approx \eta \left((a - a_\mathrm{p})^2 + \delta^2\right), \tag{11.30}$$

where $\delta \to 0$ as $Q \to 0$ and $\eta$ is some positive constant. Thus the integral for the $K(a)$ function in Eq. (6.22) becomes dominated by a "resonant peak":

$$-\int \frac{dx}{x^2} \frac{1}{\eta \left((x - a_\mathrm{p})^2 + \delta^2\right)} \approx -\frac{1}{a_\mathrm{p}^2 \eta} \frac{\pi}{\delta} + \text{finite}. \tag{11.31}$$

Therefore, in the $Q \to 0$ limit (where $\rho_1(Q) = K(a) - r_1$ tends to $-\infty$), the $\delta$ parameter vanishes $\propto 1/(-\ln Q)$. (Appendix 11.B discusses the approach to the $Q \to 0$ limit in more detail.)

The integrals $I_j(a)$ of Eq. (7.10) are also dominated by a huge peak in their integrands around $a_\mathrm{p}$:

$$I_j(a) \approx \int dx \frac{x^{j-2}}{(\eta \left((x - a_\mathrm{p})^2 + \delta^2\right))^2} \approx \frac{a_\mathrm{p}^{j-2}}{\eta^2} \frac{\pi}{2\delta^3}. \tag{11.32}$$

One can thus see that the $B_j$ and $H_j$ functions, diverge like $1/\delta^3$ as $\delta \to 0$. Note that the $B(a)/a^{j-1}$ factor in Eq. (7.9) will involve the limiting value of $a$, which is $a^\star$ and not $a_\mathrm{p}$. Although the $B_j$'s and $H_j$'s diverge, the optimized $r_m^\star$ coefficients are finite, because the $1/\delta^3$ factors cancel out, as does $\eta$, in Eq. (9.13).

**Since** the $B_j$'s diverge, the $c_k$ optimization equation $\mathcal{S} = \frac{1}{B_k(a)}$, Eq. (8.23), means that $\partial \mathcal{R}/\partial a \equiv \mathrm{P} a^{\mathrm{P}-1} \mathcal{S}$ vanishes at the unfixed point $a = a^\star$. Thus the second term in the $\tau$ optimization equation:

$$\frac{\partial \mathcal{R}}{\partial \tau} = \left.\frac{\partial \mathcal{R}}{\partial \tau}\right|_a + \frac{\beta(a)}{b} \frac{\partial \mathcal{R}}{\partial a} = 0$$

vanishes in both the fixed-point case, where $\beta(a) = 0$, and in the unfixed-point case, where $\partial \mathcal{R}/\partial a = 0$. This corresponds to the scale dimension of $\mathcal{R}$ vanishing when $\mathcal{R}$ tends to a constant. (See also Exercise 11.4.)

Instead of Eq. (11.17) of the fixed-point case, the formula for the $s_m^\star$ coefficients can be written as (see Exercise 11.5)

$$s_m^\star a_{\mathrm p}^m = \frac{1}{(k-1)} \left[ \sum_{j=0}^m (k-m-j-1) c_j^\star a_{\mathrm p}^j \right.$$

$$\left. - \left(\frac{a_{\mathrm p}}{a^\star}\right) \sum_{j=0}^{m-1} (k-m-j) c_j^\star a_{\mathrm p}^j \right], \tag{11.33}$$

where $s_m \equiv \left(\frac{\mathrm P + m}{\mathrm P}\right) r_m$. The pinch point $a_{\mathrm p}$ is where the limiting form of the $B(a)$ function touches the $a$-axis (see Fig. 11.6) and hence satisfies the two equations

$$B^\star(a)\Big|_{a=a_{\mathrm p}} = \sum_{j=0}^k c_j^\star a_{\mathrm p}^j = 0 \quad \text{and} \quad \frac{dB^\star}{da}\Big|_{a=a_{\mathrm p}} = \sum_{j=1}^k j c_j^\star a_{\mathrm p}^{j-1} = 0. \tag{11.34}$$

Using these equations one may write an alternative form of $s_m$ formula that is more convenient when $m$ is large:

$$s_m^\star a_{\mathrm p}^m = -\frac{1}{(k-1)} \left[ \sum_{j=m+1}^k (k-m-j-1) c_j^\star a_{\mathrm p}^j \right.$$

$$\left. - \left(\frac{a_{\mathrm p}}{a^\star}\right) \sum_{j=m}^k (k-m-j) c_j^\star a_{\mathrm p}^j \right]. \tag{11.35}$$

There are various ways of proceeding to combine the $r_m$'s, the pinch-point conditions, and the $\rho_j$ definitions to determine $a_{\mathrm p}, a^\star$ in terms of invariants. One general strategy is to use Eq. (11.33) for all the $r_m$'s except for $r_k$, for which Eq. (11.35) is simpler. Then, substituting in the $\rho_2, \ldots, \rho_k$ definitions one may solve, successively, for $c_2, \ldots, c_k$ in terms of $a_{\mathrm p}, a^\star$ and the invariants. Substituting in the pinch-point conditions in the form

$$\sum_{j=0}^{k-1} (k-j) c_j^\star a_{\mathrm p}{}^j = 0 \quad \text{and} \quad \sum_{j=1}^k j c_j^\star a_{\mathrm p}^{j-1} = 0 \tag{11.36}$$

(so that the highest power of $a_p$ involved is $a_p^{k-1}$) gives a pair of equations that determine $a_p$ and $a^\star$ in terms of the invariants alone.

We do not go further with the general case, but now specialize to low orders and the P $= 1$ case.

**At third order** the pinch mechanism can occur, but only under certain restrictive conditions (which are never satisfied in the $e^+e^-$ QCD case, but for other physical quantities, or other theories, the possibility could arise.) At third order the $B(a) \equiv 1 + ca + c_2a^2$ function can obviously be rewritten in the form

$$B(a) = \eta \left( (a - a_p)^2 + \delta^2 \right), \tag{11.37}$$

with

$$\eta = c_2, \quad -2a_p\eta = c, \quad \eta(a_p^2 + \delta^2) = 1. \tag{11.38}$$

If $\eta = c_2$ is positive and $c$ is negative, $B(a)$ has a minimum at a positive $a_p = -c/(2c_2)$ that can become a pinch point if the evolution of the optimized $c_2$ coefficient results in $\delta$ tending to zero as $Q \to 0$. From Eqs. (11.33) and (11.35), respectively, one finds

$$r_1^\star = -\frac{1 + ca^\star}{2a^\star}, \quad r_2^\star = -\frac{2}{3}\frac{a_p}{a^\star}c_2^\star. \tag{11.39}$$

From Eq. (11.38) with $\delta \to 0$ (or, equivalently, from Eq.(11.34)) one obtains

$$a_p = -\frac{2}{c}, \quad c_2^\star = \frac{c^2}{4}. \tag{11.40}$$

(Using these one may rewrite $r_2^\star$ as $-\frac{c}{3a^\star}$.) Substituting in the definition of $\rho_2$ yields a quadratic equation for $a^\star$:

$$1 - \frac{4}{3}ca^\star + 4\left(\rho_2 - \frac{c^2}{2}\right)a^{\star 2} = 0. \tag{11.41}$$

The infrared limit of $\mathcal{R}$ can be written, using Eq. (11.39), as

$$\mathcal{R}^\star = \frac{1}{6}a^\star(3 - ca^\star). \tag{11.42}$$

As noted above, the pinch mechanism requires $c$ to be negative, so Eq. (11.41) will only have a positive root if $\rho_2 - \frac{c^2}{2}$ is negative. Finally, the pinch mechanism requires $a^\star > a_p$ which requires $\rho_2/c^2 > 13/48$ (and for smaller $\rho_2$'s the fixed-point mechanism takes over). In summary, the pinch mechanism can operate at third order if and only if

$$c < 0 \quad \text{and} \quad \frac{13}{48} < \frac{\rho_2}{c^2} < \frac{1}{2}. \tag{11.43}$$

**At fourth order** $(k = 3)$ one finds

$$r_1^\star = \frac{(a^\star - 2a_p)}{4a^\star a_p},$$

$$r_2^\star = -\frac{1}{6a^\star a_p}(1 + ca^\star + 2c_2^\star a^\star a_p), \tag{11.44}$$

$$r_3^\star = -\frac{3}{8}\frac{a_p}{a^\star}c_3^\star.$$

Following the strategy outlined above (but see Exercise 11.6 for an alternative method) we use the $\rho_2$ and $\rho_3$ expressions to solve for $c_2^\star$ and $c_3^\star$:

$$c_2^\star = \frac{3a^{\star 2} - 4a^\star a_p(1 - 5ca^\star) + 12a_p^2(1 - 2ca^\star + 4\rho_2 a^{\star 2})}{32a^{\star 2}a_p^2}, \tag{11.45}$$

$$c_3^\star = \frac{-a^{\star 3} + a^{\star 2}a_p(2 - 5ca^\star) - 4a^\star a_p^2(1 - 3ca^\star) + 4a_p^3(2 - ca^\star + 4\rho_3 a^{\star 3})}{4a^{\star 2}(4a^\star - 3a_p)a_p^3}. \tag{11.46}$$

Substituting into $3 + 2ca_p + c_2 a_p^2 = 0$ and $c + 2c_2 a_p + 3c_3 a_p^2 = 0$ (which are equivalent to $B'(a_p) = B(a_p) = 0$) yields

$$99a^{\star 2} - 4(1 - 21ca^\star)a^\star a_p + 12(1 - 2ca^\star + 4\rho_2 a^{\star 2})a_p^2 = 0, \quad (11.47)$$

$$-a^{\star 2}(1 - 84ca^\star) + 12(1 - 5ca^\star + 16\rho_2 a^{\star 2})a^\star a_p$$
$$+ 12(5 + 2ca^\star - 12\rho_2 a^{\star 2} + 16\rho_3 a^{\star 3})a_p^2 = 0. \tag{11.48}$$

Eliminating $a_p^2$ between these equations gives $a_p$ in terms of $a^\star$ as

$$a_p = \frac{a^\star(62 + 14ca^\star + (21c^2 - 148\rho_2)a^{\star 2} - 6(7\rho_2 c - 33\rho_3)a^{\star 3})}{2(2 - 31ca^\star + 3(4\rho_2 - c^2)a^{\star 2} + 4(6\rho_2 c + \rho_3)a^{\star 3} + 12(4\rho_2^2 - 7\rho_3 c)a^{\star 4})}.$$

(11.49)

Substituting back into either equation yields a 6th-order polynomial equation that determines $a^\star$ in terms of the invariants:

$$\begin{aligned}
0 = {} & 11680 + 2224ca^\star + 3(5997c^2 - 17264\rho_2)a^{\star 2} \\
& + 2(8235c^3 - 33624\rho_2 c + 36976\rho_3)a^{\star 3} \\
& + 18(147c^4 - 2184\rho_2 c^2 + 4640\rho_2^2 + 502\rho_3 c)a^{\star 4} \\
& + 324(-49\rho_2 c^3 + 152\rho_2^2 c + 161\rho_3 c^2 - 528\rho_3\rho_2)a^{\star 5} \\
& + 108(-147\rho_2^2 c^2 + 528\rho_2^3 + 343\rho_3 c^3 - 1386\rho_3\rho_2 c + 1089\rho_3^2)a^{\star 6}.
\end{aligned}$$

(11.50)

The final result for the infrared limit of $\mathcal{R}$ at fourth order can be expressed as

$$\mathcal{R}^\star = \frac{a^\star(2a^\star a_p + 12a_p{}^2 + 3a^{\star 2}(2 + ca_p))}{24a_p{}^2}.$$

(11.51)

Note that $a^\star \geq a_p$ is needed for this solution to be relevant. One can check that the special case $a^\star = a_p$ is indeed the boundary between the pinch mechanism and the fixed-point mechanism, and corresponds to where $\gamma^\star = 0$. From such an analysis one can determine the precise (non-integer) $n_f$ values where the switchover from one mechanism to the other takes place.

## 11.7. Numerical Results for $\mathcal{R}_{e^+e^-}$

In this section, we present the numerical results from OPT in the infrared limit for $\mathcal{R}_{e^+e^-}$ in QCD with $n_f$ flavours of massless quarks. We consider integer $n_f$ from 0 up to 16. (At higher $n_f$'s the theory would not be asymptotically free.) The inputs to the numerical calculations are collected in Table 11.1, which lists the RS-invariant quantities $c$, $\rho_2$, $\rho_3$. These values are obtained from the formulas quoted in Appendix 10.A. The Feynman-diagram calculations used

Table 11.1.   Values of the invariants for $\mathcal{R}_{e^+e^-}$ (with $\sum q_i = 0$) found from the formulas quoted in Appendix 10.A, obtained from the Feynman-diagram calculations cited in the Bibliography.

| $n_f$ | $c$ | $\rho_2$ | $\rho_3$ |
|---|---|---|---|
| 0 | 2.31818 | −7.066723 | −184.37823 |
| 1 | 2.16129 | −8.397865 | −147.27522 |
| 2 | 1.98276 | −9.842342 | −113.85683 |
| 3 | 1.77778 | −11.417129 | −83.83139 |
| 4 | 1.54 | −13.144635 | −56.87785 |
| 5 | 1.26087 | −15.055062 | −32.63303 |
| 6 | 0.928571 | −17.190118 | −10.67155 |
| 7 | 0.526316 | −19.609073 | 9.52688 |
| 8 | 0.029412 | −22.399086 | 28.63336 |
| 9 | −0.6 | −25.693806 | 47.57023 |
| 10 | −1.42308 | −29.709122 | 67.70445 |
| 11 | −2.54545 | −34.817937 | 91.25169 |
| 12 | −4.16667 | −41.724622 | 122.21944 |
| 13 | 6.71429 | −51.938541 | 168.96670 |
| 14 | −11.3 | −69.384046 | 252.90695 |
| 15 | −22.0 | −108.450422 | 452.02327 |
| 16 | −75.5 | −298.641242 | 1466.56390 |

are cited in the Bibliography. In order to simplify the comparison of different $n_f$ cases — and to avoid having to assign specific electric charges to fictitious extra quarks — we have chosen to drop the terms proportional to $\sum q_i$. Those terms were included in the examples in Chapter 10, but they make very little difference to the infrared limit in the phenomenologically relevant case of $n_f = 2$.

**At third order** only the $c$ and $\rho_2$ information is used. The OPT results are given in Table 11.2. The quoted error estimate on $\mathcal{R}$ corresponds to the last term, $r_2 a^3$, of the truncated perturbation series, evaluated in the optimized RS. Also listed are values of the fixed-point couplant and the exponent $\gamma^*$. The $\mathcal{R}^*$ results are plotted in Fig. 11.7. At low $n_f$'s the uncertainty is very large, about 100%, but at larger $n_f$'s the results become increasingly precise.

**At fourth order** the results are given in Table 11.3. The quoted error estimate on $\mathcal{R}$ corresponds to the last term, $r_3 a^4$, of the optimized perturbation series. Also listed are values of the fixed-point, or the unfixed-point and pinch-point. The fixed-point

Table 11.2. Third-order OPT results for the infrared-limit of $\mathcal{R}_{e^+e^-}$ (with $\sum q_i = 0$) for different $n_f$ values. The last column gives values for the critical exponents $\gamma^*$ which characterize the power-law approach of $\mathcal{R}$ to its fixed-point limit; $\mathcal{R}^* - \mathcal{R} \propto Q^{\gamma^*}$.

| $n_f$ | $a^*$ | $\mathcal{R}^*$ | $\gamma^*$ |
|---|---|---|---|
| 0 | 0.31328 | $0.40 \pm 0.36$ | 4.70 |
| 1 | 0.28746 | $0.37 \pm 0.31$ | 3.89 |
| 2 | 0.26466 | $0.33 \pm 0.27$ | 3.23 |
| 3 | 0.24422 | $0.30 \pm 0.23$ | 2.68 |
| 4 | 0.22559 | $0.28 \pm 0.20$ | 2.21 |
| 5 | 0.20837 | $0.25 \pm 0.18$ | 1.81 |
| 6 | 0.19218 | $0.23 \pm 0.15$ | 1.47 |
| 7 | 0.17669 | $0.21 \pm 0.13$ | 1.17 |
| 8 | 0.16160 | $0.19 \pm 0.11$ | 0.92 |
| 9 | 0.14658 | $0.17 \pm 0.09$ | 0.70 |
| 10 | 0.13132 | $0.15 \pm 0.07$ | 0.52 |
| 11 | 0.11542 | $0.13 \pm 0.05$ | 0.36 |
| 12 | 0.09846 | $0.11 \pm 0.04$ | 0.23 |
| 13 | 0.07998 | $0.086 \pm 0.025$ | 0.14 |
| 14 | 0.05954 | $0.063 \pm 0.013$ | 0.066 |
| 15 | 0.03691 | $0.038 \pm 0.005$ | 0.022 |
| 16 | 0.01249 | $0.0126 \pm 0.0005$ | 0.002 |

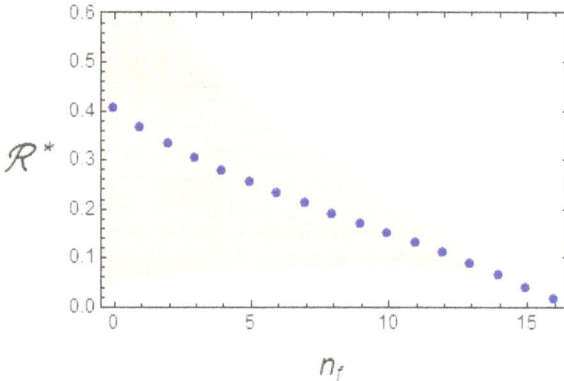

Fig. 11.7. Infrared limiting values of $\mathcal{R}_{e^+e^-}$ (with $\sum q_i = 0$) in third-order OPT as a function of $n_f$. The shaded region indicates the error estimate.

Table 11.3. Infrared-limit results for $\mathcal{R}_{e^+e^-}$ (with $\sum q_i = 0$) in OPT at fourth order for different $n_f$ values. For $n_f = 0, \ldots, 6$ and $n_f = 16$ the limit is governed by a fixed point at $a^*$: For $n_f = 7, \ldots, 15$ it arises from the pinch mechanism, with an "unfixed point" at $a^*$ and a "pinch point" at $a_{\mathrm{p}}$. (The $a^*$ equation has solutions outside this range, giving the values in parentheses, but these violate the $a^* > a_{\mathrm{p}}$ requirement. Also, the fixed-point equation has a solution for $n_f = 15$, but one that violates the $\gamma^* \geq 0$ requirement.) The last column gives values for the critical exponents $\gamma^*$ which characterize the power-law approach of $\mathcal{R}$ to its fixed-point limit; $\mathcal{R}^* - \mathcal{R} \propto Q^{\gamma^*}$. In the unfixed-point case, one finds instead $\mathcal{R}^* - \mathcal{R} \propto 1/(\ln Q)^2$ so that, in a sense, $\gamma^*$ vanishes.

| $n_f$ | $a^*$ | $a^*$ | $a_{\mathrm{p}}$ | $\mathcal{R}^*$ | $\gamma^*$ |
|---|---|---|---|---|---|
| 0 | 0.158279 | (0.1334) | (2.50) | $0.164 \pm 0.083$ | 3.28 |
| 1 | 0.168688 | (0.1465) | (1.20) | $0.174 \pm 0.083$ | 3.20 |
| 2 | 0.180844 | (0.1633) | (0.832) | $0.185 \pm 0.080$ | 3.09 |
| 3 | 0.195462 | (0.1857) | (0.651) | $0.199 \pm 0.073$ | 2.94 |
| 4 | 0.213910 | (0.2162) | (0.540) | $0.214 \pm 0.059$ | 2.73 |
| 5 | 0.239369 | (0.2588) | (0.462) | $0.235 \pm 0.028$ | 2.40 |
| 6 | 0.282493 | (0.3164) | (0.402) | $0.266 \pm 0.051$ | 1.76 |
| 7 | — | 0.383293 | 0.3525 | $0.35 \pm 0.37$ | 0 |
| 8 | — | 0.432267 | 0.3094 | $0.48 \pm 0.64$ | 0 |
| 9 | — | 0.429519 | 0.2702 | $0.52 \pm 0.75$ | 0 |
| 10 | — | 0.376034 | 0.2341 | $0.44 \pm 0.61$ | 0 |
| 11 | — | 0.301883 | 0.2001 | $0.32 \pm 0.38$ | 0 |
| 12 | — | 0.229746 | 0.1673 | $0.21 \pm 0.21$ | 0 |
| 13 | — | 0.166832 | 0.1346 | $0.14 \pm 0.11$ | 0 |
| 14 | — | 0.112784 | 0.1007 | $0.08 \pm 0.05$ | 0 |
| 15 | (0.0674) | 0.065248 | 0.0642 | $0.043 \pm 0.015$ | 0 |
| 16 | 0.020058 | (0.0215) | (0.0228) | $0.013 \pm 0.001$ | 0.001 |

mechanism operates for $n_f < 6.727$, then the pinch mechanism takes over until $n_f = 15.191$, when the fixed-point mechanism returns and operates until $n_f = 16\frac{1}{2}$ when $a^* \to 0$.

Table 11.3 also gives the values of the exponent $\gamma^*$ in the power law $\mathcal{R}^* - \mathcal{R} \propto Q^{\gamma^*}$ that applies at a fixed point. At fourth order it is given by

$$\gamma^* = ba^*(3 + 2ca^* + c_2{}^* a^{*2}). \tag{11.52}$$

Note that $\gamma^*$ is around 2 or 3 for $0 \leq n_f \leq 6$, so the resulting low-$Q$ behaviour (Fig. 11.3) is appropriately described as "freezing"

Table 11.4. Terms in the optimized $\beta$-function and $\mathcal{R}$ series in the infrared limit ($a = a^*$ or $a^\star$, as appropriate).

| $n_f$ | $ca$ | $c_2 a^2$ | $c_3 a^3$ | $r_1 a$ | $r_2 a^2$ | $r_3 a^3$ |
|---|---|---|---|---|---|---|
| 0 | 0.36692 | 0.032328 | −1.39925 | −0.25 | −0.238596 | 0.524718 |
| 1 | 0.364583 | −0.060272 | −1.30431 | −0.25 | −0.20734 | 0.489117 |
| 2 | 0.35857 | −0.183906 | −1.17466 | −0.25 | −0.165126 | 0.440499 |
| 3 | 0.347489 | −0.353982 | −0.993507 | −0.25 | −0.106587 | 0.372565 |
| 4 | 0.329421 | −0.599622 | −0.729799 | −0.25 | −0.021696 | 0.273674 |
| 5 | 0.301813 | −0.987899 | −0.313913 | −0.25 | 0.112331 | 0.117717 |
| 6 | 0.262315 | −1.74675 | 0.484438 | −0.25 | 0.371865 | −0.181664 |
| 7 | 0.201733 | −3.98554 | 2.80955 | −0.228168 | 1.11073 | −0.968966 |
| 8 | 0.012714 | −5.89311 | 5.48146 | −0.150668 | 1.72852 | −1.47106 |
| 9 | −0.257711 | −6.75979 | 7.37992 | −0.102638 | 2.05663 | −1.74115 |
| 10 | −0.535126 | −6.02116 | 6.90794 | −0.098434 | 1.8826 | −1.61273 |
| 11 | −0.768429 | −4.50798 | 5.11621 | −0.122888 | 1.44444 | −1.27189 |
| 12 | −0.957273 | −3.02694 | 3.3723 | −0.156741 | 0.999204 | −0.921032 |
| 13 | −1.12016 | −1.83192 | 2.08731 | −0.190139 | 0.635461 | −0.631526 |
| 14 | −1.27446 | −0.907359 | 1.20999 | −0.220064 | 0.353673 | −0.405223 |
| 15 | −1.43545 | −0.183663 | 0.619584 | −0.245711 | 0.135041 | −0.228425 |
| 16 | −1.51438 | 0.352822 | 0.161556 | −0.25 | −0.031878 | −0.060584 |

of the couplant. However, when $\gamma^*$ is very small one sees instead a "spiking" at $Q \to 0$. Thus, the $n_f = 16$ case, although it has a fixed point, is qualitatively similar to the "logarithmic spiking," $\mathcal{R}^\star - \mathcal{R} \propto 1/(\ln Q)^2$, of the unfixed-point case.

Table 11.4 gives the optimized coefficients, weighted by the appropriate power of $\bar{a}$, in both the $\beta$ function and $\mathcal{R}$ series. This information shows the behaviour of the truncated series for both $\mathcal{R}$ and $B(a)$ — which is, at best, only marginally satisfactory: Clearly, by going to the $Q \to 0$ limit we are pushing low-order perturbation theory well beyond its comfort zone. Nevertheless, all things considered, we believe that the results are credible within the large uncertainties quoted in Table 11.3 and illustrated in Figs. 11.3 and 11.4. In particular, we believe that the dramatic $Q \to 0$ spike produced by the pinch mechanism is real; the very large error estimate just cautions that the height of the spike is very uncertain; it might be somewhat smaller, or it might well be considerably bigger.

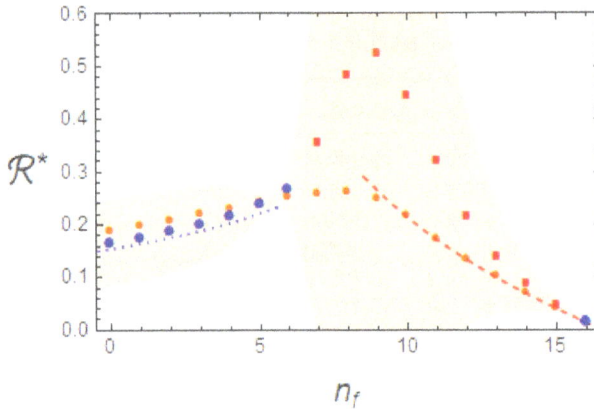

Fig. 11.8. Infrared limiting values of $\mathcal{R}_{e^+e^-}$ (with $\sum q_i = 0$) in fourth order OPT as a function of $n_f$. The OPT points are shown as blue circles when they arise from a fixed point and as red squares when they arise from an "unfixed point." The shaded region indicates the error estimate on $\mathcal{R}^*$. (The huge uncertainties for $7 \lesssim n_f \lesssim 12$ are because the height of the "spike" is so uncertain; error estimates at low, but finite $Q$ values are much more modest.) The smaller orange points are the corresponding results in the EC scheme. The fourth-order Banks–Zaks expansion about $n_f = 16.5$ is shown by the pink dashed curve. The dotted blue curve represents $\mathcal{R}^* = 0.84/b$, the asymptotic behaviour of fourth-order OPT for large-$b$.

Figure 11.8 plots the infrared limiting $\mathcal{R}$ values against $n_f$. The large "bump" around $n_f \approx 9$ is where the pinch mechanism produces really dramatic spiking of $\mathcal{R}$ as $Q \to 0$, as seen in Fig. 11.4 for $n_f = 8$. If, instead of $\mathcal{R}^*$, we had plotted $\mathcal{R}(Q)$ for some low, but finite $Q$ — say around $\frac{1}{2}\tilde{\Lambda}_\mathcal{R}$ — the bump would not have appeared and the points would have been close to the smaller, orange points. Moreover, the error estimates would have been much less.

Those smaller, orange points are the infrared-limiting results in the FAC/EC scheme, defined such that all the $r_m$ coefficients vanish, giving $\mathcal{R} = a_{\mathrm{EC}}(1 + 0 + 0 + \cdots)$. The EC $\beta$ function's coefficients then coincide with the $\rho_n$ invariants (and so can be read off from Table 11.1). Since those coefficients do not evolve with $Q$, the infrared limit in EC is simply obtained by finding the fixed point of the EC $\beta$ function. Many authors have observed that, at low orders, the EC results are very similar to those of OPT. That observation holds true at low $n_f$ and close to $n_f = 16\frac{1}{2}$. It also holds in the range $7 \lesssim n_f \lesssim 13$ at energies $Q \gtrsim \frac{1}{2}\tilde{\Lambda}_{\mathcal{R}}$. However, there is a notable difference at the very lowest energies: The EC scheme does not see the extreme spiking at $Q = 0$. While it is still true, because the error estimates (see Table 11.3) are so large in this region, that OPT and EC infrared results agree within the error estimate, it is fair to say that the presence or absence of the spike is a qualitative difference in the predictions of the two schemes.

## 11.8. Discussion

We now briefly discuss the implications of these results for QCD. The abrupt change around $n_f = 7$ (and clearly associated with the change of sign of $c$ occurring at $n_f = 8\frac{1}{19}$) seems indicative of a phase transition. For $n_f \leq 6$ the phase is presumably the one we are familiar with in the real world; colour is confined and chiral symmetry is broken, with the associated Goldstone bosons (pions) being massless when quark masses are neglected. Vector mesons ($\rho$'s, etc.) have masses of order $\tilde{\Lambda}$ and their resonant contribution dominates $e^+e^- \to$ *hadrons* at low energies. Although the actual $R_{e^+e^-} \propto 1 + \mathcal{R}$ is very different from the smooth perturbative prediction, the two agree well after Poggio–Quinn–Weinberg smearing is applied to both (see Ref. [15] and other references in the Bibliography).

An interesting possibility is that the low-$n_f$ theories can be understood through the "large-$b$ approximation." That approach can be viewed as an extrapolation from $n_f = -\infty$ (remember that $b = (33 - 2n_f)/6$ must be positive for asymptotic freedom). We shall not discuss the large-$b$ approximation here, since it involves highly technical all-orders Feynman-diagram issues. The technical subtleties in the infrared region are particularly difficult, but there are good reasons to expect that $\mathcal{R}^* \propto 1/b$, where the constant of proportionality is not universal but should, in principle, be calculable for any specific physical quantity. The large-$b$ limit of fourth-order OPT (see Exercise 11.7) is $\mathcal{R}^* \approx 0.84/b$, which is shown by the dotted blue line in Fig. 11.8. Intriguingly, it well describes the OPT results up to $n_f = 6$.

For $n_f > 7$ the infrared limit is described by an *unfixed* point. The effective low-energy theory here seems to be a *renormalizable* theory with the energy scale $Q$ appearing only in logarithms. The extreme spiking of $\mathcal{R}$ as $Q \to 0$ (Fig. 11.4), if viewed as a resonant peak in the vector channel, hints that massless vector bosons are now present. These might be the gluons of an unconfined phase, or they might be massless, colourless vector mesons of a confined phase, perhaps with unbroken chiral symmetry. This phase presumably persists until $n_f = 16\frac{1}{2}$ (beyond which asymptotic freedom is lost). Although the OPT results switch back to a fixed-point limit for $n_f = 16$, there is hardly any qualitative difference between the extreme (logarithmic) spiking of the unfixed-point case and the very strong (fractional power-law) spiking of a fixed-point with a very small $\gamma^*$. Note that the theory with 16 flavours (or 16.4999, for that matter) is not *exactly* scale and conformal invariant. Moreover, the phrase "approximately conformal invariant" needs to be used with care. While there is a huge range of $Q$ over which $\mathcal{R}$ is nearly constant (at a value about 0.78 of its infrared limit, as will be discussed in Chapter 13), it does fall to zero (very slowly) as $Q \to \infty$ and it does rise (very abruptly) as $Q \to 0$.

The approach to $n_f = 16\frac{1}{2}$ is very interesting. Indeed, one can make an expansion about that limit, as first noted by Banks and Zaks. This Banks–Zaks (BZ) expansion is essentially a small-$b$ expansion. We return to discuss it, and to consider its regime as a playground to explore some aspects of OPT at arbitrarily high orders, in Chapter 13.

Before concluding this chapter we remark that it is an open question whether the fixed-point and pinch mechanisms are the only two ways that a finite infrared limit can occur in OPT. Possibly, there might be still more exotic possibilities, especially in higher orders, for certain ranges of values of the $c$ and $\rho_j$ invariants.

## Appendix 11.A: Effective Exponents and the Slope of $\beta(a)$

According to the usual lore, the *slope* of the $\beta$ function at a fixed point, $\dot{\beta}^*$, is scheme invariant. Moreover, it supposedly can be identified with the critical exponent $\gamma^*$ that governs the power-law approach of any physical quantity $\mathcal{R}$ to its fixed-point value:

$$(\mathcal{R}^* - \mathcal{R}) \propto Q^{\gamma^*} \tag{11A.1}$$

as $Q \to 0$. These statements are sort-of true, but not quite.

**Indeed, a stark contradiction** arises if the slope of the $\beta$ function at the fixed point, is taken to be a scheme-invariant quantity. See Exercise 11.2.

The formal issue can be resolved by defining an "effective exponent" $\gamma(Q)$ associated with a specific physical quantity $\mathcal{R}$. It is related to the slope of the $\beta$ function but has an extra term that is crucial for its RS invariance. The following discussion will be at the formal level. In the next appendix, we comment on issues arising from the need to approximate and discuss the approach to $Q \to 0$ in OPT.

As usual, we consider a dimensionless physical quantity $\mathcal{R}$ with the perturbation expansion

$$\mathcal{R} = a^{\mathrm{P}}(1 + r_1 a + \cdots). \tag{11A.2}$$

Since $\mathcal{R}$ is a physical quantity and $Q$ is a physical parameter, the successive logarithmic derivatives of $\mathcal{R}$:

$$\mathcal{R}_{[n+1]} \equiv Q \frac{d\mathcal{R}_{[n]}}{dQ} \tag{11A.3}$$

for $n = 1, 2, 3, \ldots$, with $\mathcal{R}_{[1]} \equiv \mathcal{R}$, must be RS-invariant quantities, for any $Q$. In particular, the combination

$$\gamma(Q) \equiv \frac{\mathcal{R}_{[3]}}{\mathcal{R}_{[2]}} = 1 + Q \frac{d^2\mathcal{R}}{dQ^2} \bigg/ \frac{d\mathcal{R}}{dQ} \tag{11A.4}$$

is RS invariant. It is the exponent of the local-power-law form of $\mathcal{R}(Q)$ in the following sense: Take the first three terms of the Taylor expansion of $\mathcal{R}$ about $Q = Q_0$ and fit them to the power-law form

$$\mathcal{R} \approx K + CQ^{\gamma} \tag{11A.5}$$

to find

$$\mathcal{R}_0 \equiv \mathcal{R}|_{Q=Q_0} = K + CQ_0^\gamma,$$

$$\mathcal{R}_0' \equiv \left.\frac{d\mathcal{R}}{dQ}\right|_{Q=Q_0} = \gamma CQ_0^{\gamma-1}, \qquad (11\text{A}.6)$$

$$\mathcal{R}_0'' \equiv \left.\frac{d^2\mathcal{R}}{dQ^2}\right|_{Q=Q_0} = \gamma(\gamma-1)CQ_0^{\gamma-2}.$$

These algebraic equations can be inverted to find the three parameters $K, C$, and $\gamma$. (Note that $K$ is not $\mathcal{R}_0$ in general, though it is when $Q_0 \to 0$, assuming $\gamma > 0$.) In particular,

$$\gamma = 1 + Q_0 \frac{\mathcal{R}_0''}{\mathcal{R}_0'}, \qquad (11\text{A}.7)$$

which is the $\gamma(Q_0)$ of Eq. (11A.4).

**Note that** $\mathcal{R}_{[2]}/\mathcal{R}$ is $\mathcal{D}_{(\mathcal{R})}$, the scale dimension of $\mathcal{R}$, while $\gamma(Q)$ is $(Q/\mathcal{R}_{[2]})d\mathcal{R}_{[2]}/dQ$, and hence is the scale dimension of $\mathcal{R}_{[2]}$.

At high energies, where $\mathcal{R} \propto (1/\ln Q)^{\text{P}}$, one has a negative $\gamma$, but as $Q$ is lowered $\gamma$ becomes positive. As $Q \to 0$ it becomes the critical exponent $\gamma^*$ that (for any P) governs the approach of $\mathcal{R}$ to its fixed-point value $\mathcal{R}^*$:

$$(\mathcal{R}^* - \mathcal{R}) \propto Q^{\gamma^*} \quad \text{as } Q \to 0. \qquad (11\text{A}.8)$$

In the perturbative expansion of $\mathcal{R}$, in some specific RS with renormalization scale $\mu$, the only $Q$ dependence resides in the series coefficients $r_i$. For dimensional reasons, these can only depend on $Q$ through the ratio $Q/\mu$. Thus, we have

$$Q\frac{d\mathcal{R}}{dQ} = \sum_i Q\frac{dr_i}{dQ}a^{\text{P}+i} = -\sum_i \mu\frac{dr_i}{d\mu}a^{\text{P}+i} = -\left.\mu\frac{\partial\mathcal{R}}{\partial\mu}\right|_a, \qquad (11\text{A}.9)$$

where the $\mu$ partial derivative is taken holding $a$ constant. The $\mu$ RG equation says that the total $\mu$ derivative of $\mathcal{R}$ vanishes:

$$0 = \mu\frac{d\mathcal{R}}{d\mu} = \left.\mu\frac{\partial\mathcal{R}}{\partial\mu}\right|_a + \beta(a)\frac{\partial\mathcal{R}}{\partial a}, \qquad (11\text{A}.10)$$

so that one has

$$\mathcal{R}_{[2]} \equiv Q\frac{d\mathcal{R}}{dQ} = \beta(a)\frac{\partial\mathcal{R}}{\partial a}. \qquad (11\mathrm{A}.11)$$

Since $\mathcal{R}_{[2]}$ is itself a physical quantity, we can apply the same argument to it to get

$$\mathcal{R}_{[3]} = \beta(a)\frac{\partial\mathcal{R}_{[2]}}{\partial a} = \beta(a)\left(\frac{d\beta}{da}\frac{\partial\mathcal{R}}{\partial a} + \beta(a)\frac{\partial^2\mathcal{R}}{\partial a^2}\right). \qquad (11\mathrm{A}.12)$$

Dividing Eq. (11A.12) by Eq. (11A.11) yields the key result:

$$\gamma(Q) = \frac{d\beta}{da} + \beta(a)\frac{\partial^2\mathcal{R}}{\partial a^2} \Big/ \frac{\partial\mathcal{R}}{\partial a}. \qquad (11\mathrm{A}.13)$$

**We digress** briefly to recall a similar issue for the anomalous dimension of a Green's function or proper vertex, $\Gamma$, which is defined as

$$\gamma_{(\Gamma)} \equiv \frac{\mu}{\Gamma}\frac{d\Gamma}{d\mu} = \frac{1}{\Gamma}\left(\mu\left.\frac{\partial\Gamma}{\partial\mu}\right|_a + \beta(a)\frac{\partial\Gamma}{\partial a}\right).$$

It is *not* a physical quantity, but a physical quantity, the scale dimension of $\Gamma$, can be defined (echoing the discussion at the end of Sec. 2.7) as

$$\mathcal{D}_{(\Gamma)} \equiv \frac{\kappa}{\Gamma}\frac{d}{d\kappa}\Gamma(\kappa q_i, \mu, a(\mu))\Big|_{\kappa=1}.$$

(This could be written as $\frac{Q}{\Gamma}\frac{d\Gamma}{dQ}$ in terms of an overall physical scale $Q$.) The important point here is that the wavefunction-renormalization constant $Z_{(\Gamma)}$ that multiplicatively renormalizes $\Gamma$ is independent of the momentum arguments $q_i$ and cancels out in the equation above. By the dimensional argument leading to Eq. (11A.9), modified to allow for $\Gamma$ having an overall mass dimension of $D$, we see that

$$\mathcal{D}_{(\Gamma)} = D - \gamma_{(\Gamma)} + \frac{\beta(a)}{\Gamma}\frac{\partial\Gamma}{\partial a},$$

which is analogous to Eq. (11A.13).

Returning to $\gamma(Q)$, it is instructive to check directly that Eq. (11A.13) is invariant under scheme transformations. The derivatives of $\mathcal{R}$ transform as

$$\frac{\partial \mathcal{R}}{\partial a'} = \frac{\partial \mathcal{R}}{\partial a} \Big/ \frac{da'}{da},$$

$$\frac{\partial^2 \mathcal{R}}{\partial a'^2} = \frac{\partial}{\partial a}\left(\frac{\partial \mathcal{R}}{\partial a} \Big/ \frac{da'}{da}\right) \Big/ \frac{da'}{da}$$

$$= \left(\frac{\partial^2 \mathcal{R}}{\partial a^2} - \frac{\partial \mathcal{R}}{\partial a}\frac{d^2 a'}{da^2} \Big/ \frac{da'}{da}\right)\frac{1}{\left(\frac{da'}{da}\right)^2}. \qquad (11A.14)$$

Hence, the second term in Eq. (11A.13) transforms as

$$\beta'(a')\frac{\partial^2 \mathcal{R}}{\partial a'^2} \Big/ \frac{\partial \mathcal{R}}{\partial a'} = \beta(a)\frac{\partial^2 \mathcal{R}}{\partial a^2} \Big/ \frac{\partial \mathcal{R}}{\partial a} - \beta(a)\frac{d^2 a'}{da^2} \Big/ \frac{da'}{da}. \qquad (11A.15)$$

Adding this to Eq. (11.7) we see that

$$\frac{d\beta'}{da'} + \beta'(a')\frac{\partial^2 \mathcal{R}}{\partial a'^2} \Big/ \frac{\partial \mathcal{R}}{\partial u'} = \frac{d\beta}{da} + \beta(a)\frac{\partial^2 \mathcal{R}}{\partial a^2} \Big/ \frac{\partial \mathcal{R}}{\partial a}, \qquad (11A.16)$$

confirming that $\gamma(Q)$ is genuinely scheme independent.

Further insight into $\gamma(Q)$ is the following observation. Specialize to the case $\mathrm{P} = 1$ (or define $\mathcal{R}_{\mathrm{new}} = \mathcal{R}_{\mathrm{old}}^{1/\mathrm{P}}$) and consider the "effective charge" (EC) renormalization scheme defined so that $\mathcal{R} = a(1 + 0 + 0 + \cdots)$. In this scheme $\partial^2 \mathcal{R}/\partial a^2 = 0$, so Eq. (11A.13) reduces to

$$\gamma(Q) = \frac{d\beta_{\mathrm{EC}}(\mathcal{R})}{d\mathcal{R}}. \qquad (11A.17)$$

Thus $\gamma(Q)$, at any $Q$, is the slope of the EC $\beta$ function at the corresponding $\mathcal{R}$. In particular, in the infrared limit, the critical exponent $\gamma^*$ is the derivative of the EC $\beta$ function at the fixed point. Moreover, from Eq. (11A.13), we can say that $\gamma^*$ is the derivative of the $\beta$ function at the fixed point in any scheme for which $\frac{\partial \mathcal{R}}{\partial a}$ is non-zero and $\frac{\partial^2 \mathcal{R}}{\partial a^2}$ is non-singular at $a = a^*$. That includes a large class of possible RS's, but by no means is this "almost all" schemes. In general we must go back to Eq. (11A.13) and carefully consider its infrared limit. For an instance where this subtlety arises see Sec. 13.5.

An important open question concerns the "universality," or otherwise, of $\gamma^*$. Is it the same for all perturbative physical quantities $\mathcal{R}$? The question hinges on whether the EC couplants $a$ and $a'$ for two different physical quantities $\mathcal{R}$ and $\mathcal{R}'$ always have $da'/da|^*$ non-zero and $d^2a'/da^2|^*$ non-singular. Possibly yes, but it may well be that physical quantities segregate into distinct classes, each with a characteristic value of $\gamma^*$. It is this point — that $\gamma^*$ is not necessarily the same for all perturbative physical quantities of a given theory — that is perhaps the most important lesson to be drawn here.

The discussion so far has been entirely at the formal level. When we approximate $\mathcal{R}$ and $\beta(a)$ a whole set of other issues arises. These are discussed in the following appendix.

## Appendix 11.B: Approach to the $Q \to 0$ Limit

When approximating $\gamma(Q)$, or its infrared limit $\gamma^*$, the most meaningful result is just the original definition, Eq. (11A.4), with $\mathcal{R}$ replaced by its approximation. For some approximation methodologies that is the same as using the formal result, Eq. (11A.13), with the $\mathcal{R}$ and $\beta(a)$ replaced by their approximations — but that is not always the case.

Let us first consider $\mathcal{R}_{[2]} \equiv Qd\mathcal{R}/dQ$ in $(k+1)$th order. (Henceforth $\mathcal{R}$ and $\beta(a)$ should be understood as $\mathcal{R}^{(k+1)}$ and $\beta^{(k+1)}(a)$, respectively.) In fixed-RS perturbation theory, where the $Q$ dependence resides in the $Q/\mu$ dependences of the $r_i$ coefficients, the argument in Eqs. (11A.9) holds, except that the series are truncated. The next step, using the $\mu$ RG equation, will therefore not yield $\beta(a)\partial\mathcal{R}/\partial a$, but only the truncated series thereof — which does not factorize, and does not vanish when the approximated $\beta$ function vanishes. In RG-improved perturbation theory, where $\mu$ is set equal to $Q$, the $Q$ dependence is transferred to $a(Q)$, so the factorized form $\beta(a)\partial\mathcal{R}/\partial a$ is regained, with each factor being a truncated series. In general, if the RS choice evolves with $Q$ then

$$Q\frac{d\mathcal{R}}{dQ} = Q\frac{d\mathcal{R}}{dQ}\bigg|_{\text{fixed RS}} + \frac{d\mathcal{R}}{d(RS)}\frac{d(RS)}{dQ}, \qquad (11\text{B}.1)$$

where

$$\frac{d\mathcal{R}}{d(RS)}\frac{d(RS)}{dQ} \equiv \frac{\partial\mathcal{R}}{\partial\tau}\frac{d\tau}{dQ} + \sum_{j=2}^{k}\frac{\partial\mathcal{R}}{\partial c_j}\frac{dc_j}{dQ}. \qquad (11\text{B.2})$$

In OPT $\partial\mathcal{R}/\partial\tau$ and $\partial\mathcal{R}/\partial c_j$ vanish by the optimization equations. Thus, all the complications of the optimized $r_i$ and $c_j$ coefficients evolving with $Q$ can be ignored, and the fixed-RS result, Eq. (11A.9) holds. Then, by using the $\tau$ optimization equation in the role of Eq. (11A.10), one obtains the result

$$Q\frac{d\mathcal{R}}{dQ} = \beta(a)\frac{\partial\mathcal{R}}{\partial a}, \qquad (11\text{B.3})$$

where $\beta(a)$ and $\mathcal{R}$ have their optimized forms.

However, when it comes to $\mathcal{R}_{[3]}$ the same argument cannot be repeated, because the optimal scheme for $\mathcal{R}$ is not the optimal scheme for $\mathcal{R}_{[2]}$. Thus, $\gamma(Q)$ in OPT is not given by the formal expression, in general. At a fixed point, however, the naïve expression for $\gamma^*$ is valid, as we now show.

First, we note a subtlety that one must beware of. (In the following, all scheme-dependent quantities should be understood to take their optimized values.) A convenient small quantity is $\epsilon \equiv B(a)$, which goes to zero in the fixed-point limit. The difference $\mathcal{R}^* - \mathcal{R}$ will turn out to be of order $\epsilon$. However, the optimized $a$, $r_m$, and $c_j$ all have $\epsilon \ln \epsilon$ corrections as they approach their fixed-point limits. The $B$ function (which we write with a dummy argument $x$, reserving "$a$" for the optimized couplant) evolves as shown in Fig. 11.9. At some small $Q$ the $B$ function has a zero at some value $a_z$, slightly larger than $a$ itself. Clearly, $a_z - a$ is proportional to $\epsilon$, the two being related by the finite slope of $B$. However, the $B(x)$ function evolves, shifting nearly parallel to itself, so that $a^* - a_z$, and hence $a^* - a$ are of order $\epsilon \ln \epsilon$.

We now show that $a_z - a$, and hence $\epsilon$, is proportional to $Q^{\gamma^*}$, where $\gamma^*$ is indeed $\dot{\beta}^* = ba^{*2}\sigma$, the slope of the optimized $\beta$ function at the fixed point. From $\boldsymbol{\rho}_1(Q) \equiv \tau - r_1$ and the int-$\beta$ equation

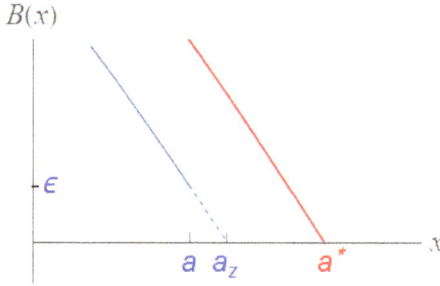

Fig. 11.9.   Sketch of the $B$ function, with dummy argument $x$, in the optimal RS as it evolves in the case of a fixed-point infrared limit. (Compare with Fig. 11.5.) Here we are showing a close-up view of the region near the fixed point. At some small $Q$ value the optimal couplant is $a$, and the value of $B(x)$ at $x = a$ is some small value $\epsilon$. The optimal $B(x)$ function at this $Q$ is shown in blue, with its continuation to $x > a$ shown dashed. It has a zero at $a_z$. At $Q = 0$ the optimal $B(x)$ function is shown in red, and the optimal infrared couplant is $a^*$, where $B$ vanishes. The subtlety is that while $a_z - a$ is of order $\epsilon$, both $a^* - a$ and $a^* - a_z$ are of order $\epsilon \ln \epsilon$. The evolving $B(x)$ is shifting nearly parallel to itself, so that the slope of $B(x)$ is almost constant throughout the region shown.

$\tau = K(a)$, we have

$$\rho_1(Q) = K(a) - r_1, \qquad (11\text{B}.4)$$

which gives

$$b\ln(Q/\tilde{\Lambda}_{\mathcal{R}}) = b \int_{[0]}^{a} \frac{dx}{\beta(x)} - r_1, \qquad (11\text{B}.5)$$

where "[0]" is shorthand for a lower limit of 0 with the appropriate infinite-constant subtraction, as in Eq. (6.17). At sufficiently low $Q$ we may approximate $\beta(x)$ by $-\gamma^*(a_z - x)$. Note that we need $a_z$ here, and not $a^*$, see Fig. 11.9. Thus,

$$b\ln Q = -\frac{b}{\gamma^*} \int_{0}^{a} \frac{dx}{(a_z - x)} + \text{const.}, \qquad (11\text{B}.6)$$

where "const." is finite as $Q \to 0$. Multiplying through by $\gamma^*/b$ gives

$$\gamma^* \ln Q = \ln(a_z - a) + \text{const.}', \qquad (11\text{B}.7)$$

and exponentiating gives

$$a_z - a \propto Q^{\gamma^*}. \qquad (11\text{B}.8)$$

Next we note that $\partial R/\partial a$ tends to a constant value — which is easily found (see Exercise 11.8), but is not needed. Thus, at low $Q$

$$Q\frac{dR}{dQ} = \beta(a)\frac{\partial R}{\partial a} \propto -(a_z - a) \propto -Q^{\gamma^*}. \tag{11B.9}$$

Integrating this with $dQ/Q$, we find the expected power law:

$$(R^* - R) \propto Q^{\gamma^*}, \tag{11B.10}$$

for $Q \to 0$. Note that the result corresponds to

$$Q\frac{dR}{dQ} \sim -\gamma^*(R^* - R), \tag{11B.11}$$

for $R$ close to $R^*$.

The case of second order is exceptional (see Exercise 11.8): There the $Q \to 0$ behaviour is *not* governed by the slope of the $\beta$ function at $a^*$. The same is true of the unfixed-point case, to which we now turn.

Approaching an unfixed point the natural small parameter is the $\delta$ of Eq. (11.30). As before, the int-$\beta$ equation and $\rho_1(Q)$ definition give Eq. (11B.5), so, using Eq. (11.31) we have

$$\delta = \frac{\pi}{ba_p^2\eta}\frac{1}{(-\ln Q/\tilde{\Lambda}_R)} \quad \text{as } Q \to 0. \tag{11B.12}$$

When we consider $R_{[2]}$ in Eq. (11B.3) it is now $\beta(a)$ that tends to a constant, while $\partial R/\partial a$ vanishes, since it is proportional to $1/B_k$, and hence to $\delta^3$. Thus, as $Q \to 0$,

$$Q\frac{dR}{dQ} \propto \frac{1}{(\ln Q)^3}. \tag{11B.13}$$

Integration with respect to $\ln Q$ then gives

$$R^\star - R = \frac{1}{b_{ir}^2}\frac{1}{(\ln Q)^2}, \tag{11B.14}$$

where the proportionality constant $1/b_{ir}^2$ can easily be found (see Exercise 11.9):

$$b_{ir} = \frac{ba_p^2}{\pi}\sqrt{\frac{(k-1)\eta}{Pa^{\star P}}}\left(\frac{a_p}{a^\star}\right)^k. \tag{11B.15}$$

It involves the constant $\eta$ of Eq. (11.30) which is one-half the second derivative of $B$ at its pinch point $a_\mathrm{p}$:

$$\eta = \frac{1}{2}\frac{d^2 B}{dx^2}\bigg|_{x=a_\mathrm{p}} = \frac{1}{2}\sum_{j=0}^{k-2}(k-j)(k-j-1)c_j a_\mathrm{p}^{j-2}, \qquad (11\mathrm{B}.16)$$

where we have used $B = B' = 0$ at $a_\mathrm{p}$ to eliminate $c_k$ and $c_{k-1}$, in a similar fashion to Eq. (11.13) for $\sigma$ in the fixed-point case.

One way to look at the result is to note that $\mathcal{R}_{[2]}$ as a function of $\mathcal{R}$ has the form

$$Q\frac{d\mathcal{R}}{dQ} \sim -2b_\mathrm{ir}(\mathcal{R}^\star - \mathcal{R})^{3/2} \qquad (11\mathrm{B}.17)$$

for $\mathcal{R}$ close to $\mathcal{R}^\star$. Thus, it has neither a simple nor a double zero, but something in between. An even more intriguing interpretation is to see the low-energy prediction as

$$\mathcal{R} = \mathcal{R}^\star - \lambda^2 (1 + O(\lambda)) \qquad (11\mathrm{B}.18)$$

with $\lambda \sim 1/(-b_\mathrm{ir}\ln Q)$ viewed as the running coupling constant of some infrared effective theory whose $\beta$ function starts $b_\mathrm{ir}\lambda^2(1+O(\lambda))$.

**Exercise 11.1.** Show that the formal argument around Eqs. (11.5) and (11.6) generalizes to any P, where, in the EC scheme,

$$\mathcal{R} = a_\mathrm{EC}^\mathrm{P} \quad \text{and} \quad \beta_\mathrm{EC}(a_\mathrm{EC}) = Q\frac{da_\mathrm{EC}}{dQ}.$$

**Exercise 11.2.** Show that a stark contradiction arises if the slope of the $\beta$ function at the fixed point,

$$\dot{\beta}^* \equiv \frac{d\beta}{da}\bigg|_{a=a^*},$$

is taken to be a scheme-invariant quantity. Write the $\beta$ function as

$$\beta(a) = -ba^2 \sum_i c_i a^i,$$

with $c_0 \equiv 1$ and $c_1 \equiv c$, and the $i$ sum extends to infinity for this formal exercise. If $\dot{\beta}^* = -b\sum_i ic_i^* a^{*i+1}$ were a physical quantity then we would have

$$\frac{\partial \dot{\beta}^*}{\partial c_j^*}\Big|_{a^*} + \frac{\partial a^*}{\partial c_j^*}\frac{d\dot{\beta}^*}{da^*} = 0.$$

Using $\partial a^*/\partial c_j^* = a^{*j}/\sigma$ (see Exercise 11.3), show that this would reduce to

$$j - \sum_i i(i+1)c_i a^{*i} \Big/ \sum_i i c_i a^{*i} = 0.$$

But this equation would have to be true for all $j = 2, 3, \ldots$, which is clearly impossible since the second term is independent of $j$.

**Exercise 11.3.** Show that the result $\partial a/\partial c_j|^* = a^{*j}/\sigma$, Eq. (11.12), obtained by considering the $I_j(a)$ integrals as $a \to a^*$, is the same as $\partial a^*/\partial c_j^*$ obtained from $\frac{\partial}{\partial c_j^*}(1 + ca^* + \cdots + c_k^* a^{*k}) = 0$.

**Exercise 11.4.** The form of the optimization equations in the fixed-point and unfixed-point infrared limits is worth considering. (Because we used the formula for the $s_m$ coefficients — which solves those equations — we did not need this discussion earlier.) A simple argument is that, if the couplant $a$ tends to a constant value $a^*$ (or $a^*$) in the infrared limit then the whole $\partial \mathcal{R}/\partial \tau$ should be the same as the partial variation with $a$ held constant:

$$\frac{\partial \mathcal{R}}{\partial \tau} \to \frac{\partial \mathcal{R}}{\partial \tau}\Big|_a = \sum_{m=1}^{k} \frac{\partial r_m}{\partial \tau} a^m.$$

Thus the $\tau$ optimization equation reduces to this term set equal to zero. This is indeed true since the other term $\beta(a)\frac{\partial \mathcal{R}}{\partial a}$ vanishes: in the fixed-point case $\beta(a)$ vanishes, while in the unfixed-point case $\frac{\partial \mathcal{R}}{\partial a}$ vanishes. However, this simple argument is not valid for the other optimization equations. Why not?

In the fixed-point case define a new RS variable

$$c_2' \equiv c_2 + c_3 a^* + \cdots + c_k a^{*k-2},$$

and change variables from $\{\tau, c_2, c_3, \ldots, c_k\}$ to $\{\tau, c_2', c_3, \ldots, c_k\}$. Now the RS variations with respect to $c_j$ for $j \geq 3$ are at constant $c_2'$. Show that the simple argument is valid for the new $c_j$ equations, though not for the $c_2'$ equation. Show that this result corresponds to the fact that $\beta_j(a) - a^{*j-2}\beta_2(a)$ vanishes at $a = a^*$.

Verify that a similar result holds in the unfixed-point case holds if we define

$$c'_2 \equiv c_2 + c_3 a_{\rm p} + \cdots + c_k a_{\rm p}{}^{k-2},$$

in that case.

**Exercise 11.5.** Derive Eq. (11.33), the formula for the $s_m$ coefficients in the pinch-mechanism case. First, from the form of the $I_j$ integrals in Eq. (11.32), find the asymptotic forms of the $B_j$, and hence the $H_i$ functions, remembering that the infrared limit of $a$ is $a^\star$, which is distinct from the pinch point $a_{\rm p}$. Then substitute in the general formula, Eq. (9.13), and simplify.

**Exercise 11.6.** An alternative method for treating the pinch mechanism in fourth order is as follows. First, use the two conditions, Eq. (11.34), satisfied by the pinch point $a_{\rm p}$ to show that

$$c_2^\star = -\frac{(3 + 2ca_{\rm p})}{a_{\rm p}^2},$$

$$c_3^\star = \frac{(2 + ca_{\rm p})}{a_{\rm p}^3}.$$

Then substitute these and Eq. (11.44) into the definitions of the $\rho_2$ and $\rho_3$ invariants to obtain two equations for $a_{\rm p}$ and $a^\star$: The first is Eq. (11.47) and the second is a linear combination of it and Eq. (11.48). (Direct manipulation of these equations leads to a result for $a_{\rm p}$ that is more cumbersome than Eq. (11.49).)

**Exercise 11.7.** Using the results for the $\rho$ invariants for $\mathcal{R}_{e+e-}$ quoted in Appendix 10.5, find the large-$b$ ($n_f \to -\infty$) limit of the fourth-order OPT formulas, Eqs. (11.28), (11.29), and show that these yield $\mathcal{R}_{e+e-}^\star \sim 0.84/b$. Show also that $\gamma^\star$ tends to a finite limit of about 2.76. (Both these results extrapolate quite well as far as $n_f \sim 6$.)

**Exercise 11.8.** Near a fixed point, as discussed in Appendix 11.B, $\partial \mathcal{R}/\partial a$ tends to a constant value. Show that, in $(k+1)$th order of OPT (for $k \geq 2$)

$$\left.\frac{\partial \mathcal{R}}{\partial a}\right|^\star = \mathrm{P}a^{\star\mathrm{P}}\frac{\sigma}{(k-1)},$$

and that

$$\mathcal{R}^* - \mathcal{R} \sim \frac{\mathrm{P}a^{*\mathrm{P}}}{(k-1)} \epsilon.$$

The case of second order $(k = 1)$ is exceptional. As discussed in Sec. 11.3, for negative $c$ the $\beta$ function has a zero at $a^* = -1/c$, but the optimal $r_1$ coefficient and $\mathcal{R}$ diverge as $Q \to 0$. From Eq. (11B.3) show that

$$Q\frac{d\mathcal{R}}{dQ} = -\mathrm{P}ba^{\mathrm{P}+1},$$

and hence show that

$$\mathcal{R} \sim \mathrm{P}ba^{*\mathrm{P}+1}\left(-\ln Q\right)$$

as $Q \to 0$.

**Exercise 11.9.** Find the constant of proportionality in Eq. (11B.13) and hence obtain the result for $b_{\mathrm{ir}}$ in Eq. (11B.15). Show that, for the $\mathrm{P} = 1$ case at fourth order $(k = 3)$ it reduces to

$$b_{\mathrm{ir}} = \sqrt{2a_{\mathrm{p}}(3 + ca_{\mathrm{p}})}\left(\frac{a_{\mathrm{p}}}{a^*}\right)^2\frac{b}{\pi}.$$

Also show that, approaching an unfixed-point infrared limit,

$$\mathcal{R}^* - \mathcal{R} \sim \frac{\mathrm{P}a^{*\mathrm{P}}}{(k-1)}\eta\left(\frac{a^*}{a_{\mathrm{p}}}\right)^k\delta^2.$$

**Exercise 11.10.** Consider third order $(k = 2)$. The nature of the infrared limit depends on the two invariants $c$ and $\rho_2$. (The former depends only on the theory, while the latter depends on the specific physical quantity being considered.) The object of this exercise is to map out, in the $c, \rho_2$ plane, the regions of different infrared behaviour.

(i) First, consider the EC scheme. The fixed point, if one exists, is the smaller positive root of $1 + ca + c_2a^2 = 0$, with $c_2 = \rho_2$. Show that contours of fixed $a^*$ are straight lines in the $c, \rho_2$ plane, whose envelope gives the boundary of the fixed-point region. That boundary is the positive $c$-axis ($\rho_2 = 0$) and the half-parabola $\rho_2 = \frac{1}{4}c^2$ for $c < 0$. Above this boundary there is Landau-pole behaviour.

(ii) Repeat the analysis for OPT. It is somewhat simpler to consider the $c, \rho_2'$ plane, where $\rho_2' \equiv \rho_2 - \frac{1}{4}c^2$, since then the contours of fixed $a^*$, given by Eq. (11.22), remain straight lines. Show that the envelope of these straight lines, for $c < 0$, is $\rho_2' = \frac{1}{21}c^2$.

(iii) However, for a viable fixed-point solution $a^*$ must not only be the smallest positive root of the $a^*$ equation, it must be the smallest positive root of $B^*(x) = 1 + cx + c_2^* x^2$. At this order, with only two roots, it suffices to check that the slope parameter $\sigma$ is positive. Show that this requires $a^* < \frac{2}{(-c)}$ and hence $\rho_2' < \frac{1}{48}c^2$. The pinch mechanism operates for $\frac{1}{48}c^2 < \rho_2' < \frac{1}{4}c^2$ for negative $c$, see Eq. (11.43). There the contours of fixed $a^\star$ are given by Eq. (11.41).

# Chapter 12

# Optimization of Factorized Quantities

## 12.1. Introduction

Factorized quantities have one non-perturbative factor and another that is perturbatively calculable. The factorization itself requires arbitrary choices, which give rise to a "factorization-scheme-dependence problem" in addition to, and intertwined with, the renormalization-scheme-dependence problem. The difficulties initially appear formidable and have only recently been clarified, but when the smoke clears the resulting optimization procedure is only slightly more complicated than for purely perturbative quantities.

The prototypical case arises in deep-inelastic leptoproduction, where a high-energy lepton collides with a proton, or other hadron, exchanging a virtual photon of large virtuality $Q^2$. Neglecting power-suppressed terms, the $n$th moment, $\int_0^1 \frac{dx}{x} x^n F(x, Q)$, of the non-singlet proton structure function can be factorized into the form

$$F_n(Q) = \langle \mathcal{O}_n(M) \rangle C_n(Q, M), \qquad (12.1)$$

where $\langle \mathcal{O}_n(M) \rangle$ is an operator matrix element, $C_n$ is a coefficient function, and $M$ is some arbitrary "factorization scale." (From now on the moment index $n$ will be suppressed.)

The operator matrix element $\langle \mathcal{O}(M) \rangle$ has an $M$ dependence given by its anomalous dimension

$$\frac{M}{\langle \mathcal{O} \rangle} \frac{d\langle \mathcal{O} \rangle}{dM} \equiv \gamma_{\mathcal{O}}. \tag{12.2}$$

While $\langle \mathcal{O}(M) \rangle$ itself cannot be calculated perturbatively, its anomalous dimension, $\gamma_{\mathcal{O}}$, has a calculable perturbation series of the form

$$\gamma_{\mathcal{O}}(a) = -bga(1 + g_1 a + g_2 a^2 + \cdots). \tag{12.3}$$

The leading-order coefficient is written as $-bg$ for later convenience. While $g$ is invariant, the other coefficients, $g_1, g_2, \ldots$ are scheme-dependent. The expansion parameter here, $a = a(M)$, is the couplant in some arbitrary RS whose renormalization scale is $M$.

The coefficient function $C$ can be calculated as a perturbation series:

$$C(Q, M) = 1 + r_1 \tilde{a} + r_2 \tilde{a}^2 + \cdots, \tag{12.4}$$

where $\tilde{a}$ is the couplant of some other arbitrary RS — which can be different from the RS used to define $a$. It can have a different renormalization scale $\tilde{M}$, and different RS labels $\tilde{c}_2, \ldots$. Perhaps the easiest way to understand that the RS's for $a$ and $\tilde{a}$ can be distinct, without inconsistency, is to imagine that first both $\langle \mathcal{O} \rangle$ and $C$ are calculated in the same RS and then a substitution $\tilde{a} = a(1 + v_1 a + v_2 a^2 + \cdots)$, with arbitrary $v_1, v_2, \ldots$, is made in the result for $C$. In terms of renormalization constants, the $Z_{\mathcal{O}}$ constant needed for the renormalization of the operator $\mathcal{O}$ (which is genuinely an infinite change of normalization) must be consistent between the calculations of $C$ and $\gamma_{\mathcal{O}}$, but the reparametrization step — the substitution of $a = Z_a a_{\text{bare}}$ and $\tilde{a} = \tilde{Z}_a a_{\text{bare}}$ in the bare forms of $\gamma_{\mathcal{O}}$ and $C$, respectively — can involve distinct $Z_a$ and $\tilde{Z}_a$ renormalization constants.

Thus, what we shall call "RS/FS dependence" involves a choice of factorization scheme (FS), parametrized by $g_1, g_2, \ldots$, and two, independent, choices of RS for $a$ and $\tilde{a}$ that are labelled, respectively, by $\tau, c_2, c_3, \ldots$ and by $\tilde{\tau}, \tilde{c}_2, \tilde{c}_3, \ldots$, where

$$\tau \equiv b \ln(M/\Lambda), \quad \tilde{\tau} \equiv b \ln(\tilde{M}/\Lambda). \tag{12.5}$$

**In this chapter** we shall omit the tildes over $\Lambda$, used previously to distinguish our definition from the more conventional one. Note that, without loss of generality, we may assume that the two RP's, for $a$ and $\tilde{a}$, are defined so that their $\Lambda$ parameters are the same: If this were not the case initially, we could trivially redefine the renormalization-scale parameter in one of the schemes by a compensating factor.

## 12.2. The Form of $\langle \mathcal{O} \rangle$ and the Invariance of Its Normalization Constant

Integrating Eq. (12.2), utilizing the $\beta$-function equation

$$M \frac{\partial a}{\partial M} = \beta(a) = -ba^2(1 + ca + c_2 a^2 + \cdots), \qquad (12.6)$$

gives

$$\langle \mathcal{O} \rangle = (\text{const.}) \exp \left( \int^a dx \frac{\gamma_\mathcal{O}(x)}{\beta(x)} \right). \qquad (12.7)$$

(Note that the $M$ dependence of $\langle \mathcal{O} \rangle$ comes solely from the $M$ dependence of $a$.) The constant of integration may be written as a constant $A$ defined by

$$\langle \mathcal{O} \rangle = A \exp \left( \int_0^a dx \frac{\gamma_\mathcal{O}(x)}{\beta(x)} - \int_0^\infty dx \frac{gx}{x^2(1+cx)} \right), \qquad (12.8)$$

where, as with the definition of $\Lambda$, the lower limit of $x \to 0$ in each integral produces a divergence that cancels between the two integrals. The normalization constant $A$ is not calculable from perturbation theory, but is RS/FS invariant, as we now show.

**Theorem (Politzer and Stevenson).** *The normalization constant $A$ in Eq. (12.8) is RS/FS invariant.*

**Proof.** The proof is directly analogous to the proof of the Celmaster–Gonsalves (CG) relation (see Sec. 6.5). Let primed and unprimed quantities refer to two different schemes (both with the same value $M$ for their scale argument). Then we have

$$a = a' \left(1 + v_1 a' + \cdots \right),$$
$$\langle \mathcal{O} \rangle = \langle \mathcal{O}' \rangle \left(1 + w_1 a' + \cdots \right), \qquad (12.9)$$

where the coefficients $v_1, w_1, \ldots$ can have arbitrary values. Taking the logarithm of Eq. (12.8), after dividing through by $A$, and then subtracting the corresponding equation in the primed scheme, gives

$$\ln\left(\frac{\langle\mathcal{O}\rangle}{A}\right) - \ln\left(\frac{\langle\mathcal{O}'\rangle}{A'}\right) = \int_0^a dx\, \frac{\gamma_{\mathcal{O}}(x)}{\beta(x)} - \int_0^{a'} dx\, \frac{\gamma'_{\mathcal{O}}(x)}{\beta'(x)}. \quad (12.10)$$

The left-hand side is

$$\ln\left(\frac{A'}{A}\right) + \ln\left(\frac{\langle\mathcal{O}\rangle}{\langle\mathcal{O}'\rangle}\right) = \ln\left(\frac{A'}{A}\right) + \ln(1 + w_1 a' + \cdots)$$

$$= \ln\left(\frac{A'}{A}\right) + O(a), \quad (12.11)$$

and the right-hand side of Eq. (12.10) is easily shown to be of order $a$. (The key difference from the $\Lambda$ case is that the $\gamma_{\mathcal{O}}$ factors make the integrands less singular at the $x = 0$ endpoint; $\frac{1}{x}$ rather than $\frac{1}{x^2}$.) As in the CG argument, we may now consider the limit $M \to \infty$, if $b > 0$ (or $M \to 0$ if $b < 0$) so that all terms of order $a$ tend to zero. Thus, we see that

$$\ln\left(\frac{A'}{A}\right) = 0, \quad (12.12)$$

so that $A' = A$. Thus $A$ is the same in all schemes. As with the CG argument, the $M \to \infty$ limit is merely a convenient trick; the unwanted terms actually cancel for any $M$. One can show this explicitly by using the method of Osborn's proof of the CG relation (see Exercise 12.1). $\qquad\square$

The notation conceals the fact that $A$ depends on the moment index $n$ and on the specific hadron whose structure function one is considering. For any given hadron the set of $A_n$'s provides a scheme-independent characterization of the hadronic wavefunction information. The $A_n$'s can be fitted to the data for one experiment (say, deep-inelastic leptoproduction) and used to make predictions for another process (say, Drell–Yan). Given the $A_n$'s one can use perturbation theory to predict the structure-function moments — or, inversely to find the $A_n$'s from experimental structure-function data. One will need perturbative calculations, for each $n$, of the $r_i$ coefficients in $C$ and the $g_i$ coefficients in $\gamma_{\mathcal{O}}$, as well as the $\beta$-function coefficients.

## 12.3. Second-Order Approximation

We first discuss the second-order approximation, which corresponds to truncating the series (12.3), (12.4), and the $\beta$-function after two terms. The integrals in Eq. (12.8) become

$$\int_0^a dx \frac{-bgx(1+g_1x)}{-bx^2(1+cx)} - \int_0^\infty dx \frac{gx}{x^2(1+cx)}$$

$$= gg_1 \int_0^a dx \frac{1}{1+cx} - g \int_a^\infty dx \left( \frac{1}{x} - \frac{c}{1+cx} \right)$$

$$= g \left( \frac{g_1}{c} \ln(1+ca) + \ln|ca| - \ln(1+ca) \right), \qquad (12.13)$$

which exponentiates to

$$|ca|^g (1+ca)^{-g(1-g_1/c)}. \qquad (12.14)$$

Substituting in Eq. (12.1), one obtains the second-order approximation to $F$ as

$$F^{(2)} = A|ca|^g (1+ca)^{-g(1-g_1/c)}(1+r_1\tilde{a}). \qquad (12.15)$$

This approximant depends on RS/FS choices through three variables, $\tau$, $\tilde{\tau}$, and $g_1$. Partial differentiations of Eq. (12.15) yield

$$\frac{1}{F^{(2)}} \frac{\partial F^{(2)}}{\partial \tilde{\tau}} = \frac{1}{(1+r_1\tilde{a})} \left( -\tilde{a}^2(1+c\tilde{a})r_1 + \tilde{a} \frac{\partial r_1}{\partial \tilde{\tau}} \right), \qquad (12.16)$$

$$\frac{1}{F^{(2)}} \frac{\partial F^{(2)}}{\partial \tau} = -ga(1+g_1a) + \frac{\tilde{a}}{(1+r_1\tilde{a})} \frac{\partial r_1}{\partial \tau}, \qquad (12.17)$$

$$\frac{1}{F^{(2)}} \frac{\partial F^{(2)}}{\partial g_1} = \frac{g}{c} \ln(1+ca) + \frac{\tilde{a}}{(1+r_1\tilde{a})} \frac{\partial r_1}{\partial g_1}. \qquad (12.18)$$

Self-consistency of perturbation theory requires these variations to be of order $a^2$. Noting that $\tilde{a} = a(1+O(a))$, we see that

$$\frac{\partial r_1}{\partial \tilde{\tau}} = 0, \quad \frac{\partial r_1}{\partial \tau} = g, \quad \frac{\partial r_1}{\partial g_1} = -g, \qquad (12.19)$$

so that $r_1$ has the form

$$r_1 = g\left(\tau - g_1 - \boldsymbol{\sigma}_1(Q)\right), \tag{12.20}$$

where $\boldsymbol{\sigma}_1(Q)$ is an invariant.

Substituting Eq. (12.19) back into Eqs. (12.16)–(12.18) and equating to zero produces the optimization conditions. Since $\partial r_1/\partial\tilde{\tau}$ vanishes, the solution to the optimization equation (12.16) is simply

$$r_1^{\text{opt}} = 0. \tag{12.21}$$

The second optimization equation, from (12.17), then reduces to

$$\tilde{a} = a(1 + g_1 a), \tag{12.22}$$

and (12.18) gives

$$\ln(1 + ca) = c\tilde{a}. \tag{12.23}$$

Eliminating $\tilde{a}$ between these last two equations gives us the optimal $g_1$ in terms of $a$:

$$g_1^{\text{opt}} = \frac{\ln(1 + ca) - ca}{ca^2}. \tag{12.24}$$

Also, from the int-$\beta$ equation at second order, we have

$$\tau = \frac{1}{a} + c\ln\left|\frac{ca}{1 + ca}\right|. \tag{12.25}$$

Substituting for $\tau$ and for $g_1$ in Eq. (12.20) and equating to zero, since $r_1^{\text{opt}} = 0$, we find

$$\ln(1 + ca) - (ca)^2 \ln\left|\frac{ca}{1 + ca}\right| = ca\left(2 - a\boldsymbol{\sigma}_1(Q)\right), \tag{12.26}$$

which determines the optimized $a$ in terms of the invariant quantities $c$ and $\boldsymbol{\sigma}_1(Q)$. Substituting back in Eq. (12.24) then fixes $g_1^{\text{opt}}$. The final optimized result, from Eq. (12.15), is

$$F_{\text{opt}}^{(2)} = A|ca|^g(1 + ca)^{-g(1 - g_1^{\text{opt}}/c)}. \tag{12.27}$$

Note that the optimization condition $r_1^{\text{opt}} = 0$ means that $C_{\text{opt}} = 1$, so that all perturbative corrections are effectively exponentiated and reabsorbed into the anomalous dimension by the optimization procedure. As we shall see later, this property holds at any order, as first noted by Nakkagawa and Niégawa.

Also note that while the value of $\tilde{a}$ (and hence $\tilde{\tau}$) is determined, it is not needed to obtain the result for $F_{\text{opt}}^{(2)}$.

## 12.4. RG Equations

As discussed above the RS/FS variables are $\tau$, $c_j$, $\tilde{\tau}$, $\tilde{c}_j$, and the $g_i$ coefficients. We now write down the RG equations expressing the fact that the physical quantity $F$ is independent of all these variables. Symbolically, we have

$$\frac{1}{F}\frac{\partial F}{\partial X} = 0, \tag{12.28}$$

where $X$ stands for any of the set of variables $\{\tau, c_j, \tilde{\tau}, \tilde{c}_j, g_j\}$.

Recalling the factorized form $F = \langle \mathcal{O} \rangle C$ of Eq. (12.1), and noting that $\langle \mathcal{O} \rangle$ is manifestly independent of $\tilde{M}$, we see that

$$\frac{1}{F}\frac{\partial F}{\partial \tilde{\tau}} = \frac{1}{C}\frac{\partial C}{\partial \tilde{\tau}}. \tag{12.29}$$

The same argument applies to the $\tilde{c}_j$ derivatives, since $\langle \mathcal{O} \rangle$, while it depends on $a$ and its RS variables $\tau, c_j$, is manifestly independent of $\tilde{a}$ and its RS variables $\tilde{\tau}, \tilde{c}_j$. Thus, the first two RG equations have the familiar form

$$\left(\frac{\partial}{\partial \tilde{\tau}}\bigg|_{\tilde{a}} + \frac{\tilde{\beta}(\tilde{a})}{b}\frac{\partial}{\partial \tilde{a}}\right)C = 0, \quad \text{``}j = 1\text{''} \tag{12.30}$$

$$\left(\frac{\partial}{\partial \tilde{c}_j}\bigg|_{\tilde{a}} + \tilde{\beta}_j(\tilde{a})\frac{\partial}{\partial \tilde{a}}\right)C = 0, \quad j = 2, 3, \ldots, \tag{12.31}$$

where the first term collects dependence from the $r_i$ coefficients of $C$, while the second term collects the compensating dependence via $\tilde{a}$.

The other RG equations all take the form

$$\frac{1}{C}\frac{\partial C}{\partial X} + \frac{1}{\langle \mathcal{O} \rangle}\frac{\partial \langle \mathcal{O} \rangle}{\partial X} = 0, \tag{12.32}$$

where $X$ is any of the variables $\tau, c_j$ or $g_j$. The first term only involves dependence via the $r_i$ coefficients — indeed we are tempted to add "$|_{\tilde{a}}$" (meaning "with $\tilde{a}$ held constant") to the notation, to match Eqs. (12.30), (12.31), but it is unnecessary since $\tilde{a}$ is manifestly independent of $\tau, c_j$ and $g_j$. The second term can be evaluated as follows. In the case $X \to \tau$, we may simply use the definition of $\gamma_{\mathcal{O}}$, Eq. (12.2), to get

$$\frac{1}{\langle \mathcal{O} \rangle} \frac{\partial \langle \mathcal{O} \rangle}{\partial \tau} = \frac{\gamma_{\mathcal{O}}}{b}. \tag{12.33}$$

For $X \to c_j$ we can first write

$$\frac{1}{\langle \mathcal{O} \rangle} \frac{\partial \langle \mathcal{O} \rangle}{\partial c_j} = \frac{1}{\langle \mathcal{O} \rangle} \frac{\partial \langle \mathcal{O} \rangle}{\partial c_j}\bigg|_a + \frac{1}{\langle \mathcal{O} \rangle} \frac{d \langle \mathcal{O} \rangle}{da} \frac{\partial a}{\partial c_j}, \tag{12.34}$$

and then use Eq. (12.8) to obtain

$$\frac{1}{\langle \mathcal{O} \rangle} \frac{\partial \langle \mathcal{O} \rangle}{\partial c_j} = \int_0^a dx \, \frac{\gamma_{\mathcal{O}}(x)}{\beta(x)^2} bx^{j+2} + \frac{\gamma_{\mathcal{O}}(a)}{\beta(a)} \beta_j(a). \tag{12.35}$$

Although we return to this form later, for the present we rewrite it as

$$\frac{1}{\langle \mathcal{O} \rangle} \frac{\partial \langle \mathcal{O} \rangle}{\partial c_j} = \int_0^a dx \frac{\beta_j(x)}{\beta(x)} \gamma'_{\mathcal{O}}(x), \tag{12.36}$$

where $\gamma'_{\mathcal{O}}(x) \equiv d\gamma_{\mathcal{O}}/dx$. The equivalence to Eq. (12.35) can be shown by integrating by parts and then using the differential equation satisfied by the $\beta_j$ functions, Eq. (7.12). (See Exercise 12.2.) Finally, for $X \to g_j$ we find, from Eq. (12.8),

$$\frac{1}{\langle \mathcal{O} \rangle} \frac{\partial \langle \mathcal{O} \rangle}{\partial g_j} = -bg \int_0^a dx \frac{x^{j+1}}{\beta(x)}. \tag{12.37}$$

Thus, the RG equations, in addition to Eqs. (12.30), (12.31), are

$$\frac{1}{C} \frac{\partial C}{\partial \tau} + \frac{\gamma_{\mathcal{O}}}{b} = 0, \quad \text{"} j = 1 \text{"} \tag{12.38}$$

$$\frac{1}{C}\frac{\partial C}{\partial c_j} + \int_0^a dx \frac{\beta_j(x)}{\beta(x)}\gamma_o'(x), = 0, \quad j = 2, 3, \ldots, \qquad (12.39)$$

$$\frac{1}{C}\frac{\partial C}{\partial g_j} - bg \int_0^a dx \frac{x^{j+1}}{\beta(x)} = 0, \quad j = 1, 2, \ldots, \qquad (12.40)$$

As usual, the RG equations determine how the coefficients $r_i$ must depend on the RS/FS variables. We now rewrite the RG equations to facilitate finding these dependences. First, we use the series for $\gamma_o$ and $C$:

$$\gamma_o(a) = -bg \sum_{i=0} g_i a^{i+1}, \quad C = \sum_{i=0} r_i \tilde{a}^i, \qquad (12.41)$$

with $r_0 \equiv g_0 \equiv 1$. Second, we convert the $\beta, \beta_j$ functions to the $B, B_j$ functions of Sec. 7.2 (whose series begin $1 + \cdots$). A third simplification, concerning the lower limit of the $i$ summations, is discussed below. We obtain

$$\sum_{i=1} \frac{\partial r_i}{\partial \tilde{\tau}}\tilde{a}^i - \tilde{a}^2 \tilde{B}(\tilde{a}) \sum_{i=1} i r_i \tilde{a}^{i-1} = 0, \qquad (12.42)$$

$$\sum_{i=j+1} \frac{\partial r_i}{\partial \tilde{c}_j}\tilde{a}^i + \tilde{a}^{j+1}\frac{\tilde{B}_j(\tilde{a})}{j-1} \sum_{i=1} i r_i \tilde{a}^{i-1} = 0, \qquad (12.43)$$

$$\frac{1}{C} \sum_{i=1} \frac{\partial r_i}{\partial \tau}\tilde{a}^i - ga \sum_{i=0} g_i a^i = 0, \qquad (12.44)$$

$$\frac{1}{C} \sum_{i=j} \frac{\partial r_i}{\partial c_j}\tilde{a}^i + \frac{g}{j-1} \int_0^a dx\, x^{j-1}\frac{B_j(x)}{B(x)} \sum_{i=0}(i+1)g_i x^i = 0, \quad (12.45)$$

$$\frac{1}{C} \sum_{i=j} \frac{\partial r_i}{\partial g_j}\tilde{a}^i + g \int_0^a dx \frac{x^{j-1}}{B(x)} = 0. \qquad (12.46)$$

The $i$ summations of the $\partial r_i/\partial X$ terms inherently begin with $i = 1$, but in the $c_j$ and $g_j$ equations, where the second term starts only at order $a^j$, it is immediately evident that $r_i$ cannot depend on $c_j$ or $g_j$ for $i < j$. Thus, we may begin those $i$ summations at $i = j$. For the $\tilde{c}_j$ equation a stronger result holds, since $\partial r_i/\partial \tilde{c}_j$ must vanish for $i = j$ as well as for $i < j$. This observation is crucial for the "exponentiation theorem" proved in Sec. 12.6.

In $(k+1)$th order all the sums would go up to $i = k$ only and the equations would be satisfied, in an arbitrary RS/FS, only up to remainder terms of order $a^{k+1}$. The vanishing of all terms up to and including $a^k$ fixes the RS/FS dependence of the $r_i$ coefficients, and leads us to identify a set of invariants, $\sigma_j$, as discussed in the next section.

## 12.5. Invariants

The scheme dependences of $r_1$ were already found in Eq. (12.19) and led us to the first invariant

$$\sigma_1(Q) = \tau - g_1 - \frac{r_1}{g}. \qquad (12.47)$$

It is $Q$ dependent because $r_1$, when calculated from Feynman diagrams, will contain a term $-bg\ln(Q/M)$. As in the discussion of Sec. 7.3, one can write $\sigma_1(Q)$ as $b\ln(Q/\Lambda_F)$, where $\Lambda_F$ is a scale specific to the quantity $F$, but related in an exactly calculable way to the $\Lambda$ of some universal, reference RS.

**The earlier literature** used an "invariant" $\kappa_1$ given by

$$\kappa_1 = r_1 + gg_1 + bg\ln(Q/M).$$

It is true that $\kappa_1$ is invariant under changes of FS and renormalization scale, with the explicit $g_1$ and $M$ dependences cancelling the implicit $g_1$ and $M$ dependences of $r_1$. Where $\kappa_1$ fails to be invariant is under a change of RS/FS that leaves $g_1$ and the renormalization scale $M$ unchanged, but changes the RP, so that $a' = a(1 + v_1 a + \cdots)$, with some arbitrary $v_1$. Under such a transformation the $a^g$ factor in $\langle \mathcal{O} \rangle$, see Eq. (12.14), becomes $(a')^g = a^g(1 + gv_1 a + \cdots)$, so the coefficient $r_1$ must become $r_1' = r_1 - gv_1$ to leave $F = \langle \mathcal{O} \rangle C$ invariant. Thus, $\kappa_1' = \kappa_1 - gv_1$. Since our $\sigma_1(Q)$ is

$$\sigma_1(Q) = b\ln(Q/\Lambda) - \kappa_1/g,$$

this change in $\kappa_1$ cancels with the change from $\Lambda$ to $\Lambda'$, by the Celmaster–Gonsalves relation.

The higher invariants, $\sigma_2, \sigma_3, \ldots$, can be defined to be $Q$-independent. As with the $\rho_j$ invariants, it is convenient to define the $\sigma_j$'s so that they reduce to the $\beta$-function coefficients $c_j$ in "effective charge" schemes, defined by the RS/FS choices $g_j = 0$, $r_i = 0$. The invariants, so defined, depend on $\tau$ and $\tilde{\tau}$ only via the difference $\tilde{\tau} - \tau$ and have no dependence on $Q$ or $\Lambda$.

To find the invariants we will need the conversion between $\tilde{a}$ and $a$; either $\tilde{a} = a(1 + V_1 a + V_2 a^2 + \cdots)$ or its inverse

$$a = \tilde{a}(1 + \tilde{V}_1\tilde{a} + \tilde{V}_2\tilde{a}^2 + \cdots). \tag{12.48}$$

As discussed in Exercise 7.4 in Chapter 7, the $V_i$ or $\tilde{V}_i$ coefficients can most easily be found from the relation between the $\beta$ functions: $\tilde{\beta}(\tilde{a}) = (d\tilde{a}/da)\beta(a)$. (In fact, the calculation mirrors that for the $\rho_i$ invariants.) The first three coefficients are

$$\tilde{V}_1 = \tilde{\tau} - \tau,$$

$$\tilde{V}_2 = (\tilde{\tau} - \tau)^2 + c(\tilde{\tau} - \tau) - (\tilde{c}_2 - c_2),$$

$$\tilde{V}_3 = (\tilde{\tau} - \tau)^3 + \frac{5}{2}c(\tilde{\tau} - \tau)^2 + (-2\tilde{c}_2 + 3c_2)(\tilde{\tau} - \tau) - \frac{1}{2}(\tilde{c}_3 - c_3). \tag{12.49}$$

Note that the $\tilde{V}_i$'s do *not* only involve differences $c_j - \tilde{c}_j$. It *is* true, though, that the $V_i$ coefficients of the inverse relationship are obtained by exchanging all plain and tilde variables.

We now turn to a calculation of the invariant $\sigma_2$. Expanding Eqs. (12.42)–(12.46) in powers of $a$ and $\tilde{a}$ and using the above relations we can extract the self-consistency conditions. From the lowest-order terms we recover Eqs. (12.19) for $r_1$'s derivatives, plus confirmation that $r_1$ does not depend on the other RS/FS variables ($c_2$, $\tilde{c}_2$, or $g_2$). From the next-order terms we find

$$\frac{\partial r_2}{\partial \tilde{\tau}} = r_1, \quad \frac{\partial r_2}{\partial \tau} = g(r_1 + g_1 + \tilde{\tau} - \tau),$$

$$\frac{\partial r_2}{\partial \tilde{c}_2} = 0, \quad \frac{\partial r_2}{\partial c_2} = -\frac{g}{2}, \tag{12.50}$$

$$\frac{\partial r_2}{\partial g_1} = -g\left(r_1 - \frac{c}{2} + \tilde{\tau} - \tau\right), \quad \frac{\partial r_2}{\partial g_2} = -\frac{g}{2}.$$

Integrating each of these equations individually is easy, but combining the results consistently is a little tricky. However, it is straightforward to check our result that $r_2$ has the form:

$$r_2 = \frac{1}{2}\left(-gc_2 + gg_1c + gg_1^2 - gg_2 + 2g_1r_1 + r_1^2 + \frac{r_1^2}{g} + 2r_1(\tilde{\tau} - \tau)\right)$$
$$+ \text{const.,} \qquad (12.51)$$

where the constant is independent of all the RS/FS variables. The constant can be conveniently written as $\frac{g}{2}\sigma_2$ so that the invariant $\sigma_2$ is given by

$$\sigma_2 = c_2 + g_2 - g_1c - g_1^2 + \frac{2r_2}{g} - 2g_1\frac{r_1}{g} - \frac{r_1^2}{g^2}(1+g) - \frac{2r_1}{g}(\tilde{\tau} - \tau). \quad (12.52)$$

This reduces to $c_2$ in the "effective charge" scheme mentioned earlier.

An easier and more systematic way to calculate the $\sigma_i$ invariants is to find them as the $\rho_i$ invariants associated with the physical quantity (the scale dimension of $F$)

$$\mathcal{D} \equiv \frac{Q}{F}\frac{dF}{dQ}. \qquad (12.53)$$

The perturbation series for $\mathcal{D}$ can be found in terms of the $C$ and $\gamma_\mathcal{O}$ series in various ways. Perhaps the simplest is the following. First, note that all the $Q$ dependence resides in the coefficients of $C$. For dimensional reasons such $Q$ dependence can come only via the ratios $Q/M$ and $Q/\tilde{M}$. Thus,

$$\mathcal{D} = \frac{Q}{C}\frac{dC}{dQ} = -\frac{1}{C}\left(M\frac{dC}{dM} + \tilde{M}\frac{\partial C}{\partial \tilde{M}}\bigg|_{\tilde{a}}\right). \qquad (12.54)$$

The $M$ dependence of $C$ must cancel out with that of $\langle\mathcal{O}\rangle$ in the product $F = \langle\mathcal{O}\rangle C$, so that

$$\frac{M}{C}\frac{dC}{dM} = -\frac{M}{\langle\mathcal{O}\rangle}\frac{d\langle\mathcal{O}\rangle}{dM} = -\gamma_\mathcal{O}, \qquad (12.55)$$

while $C$ is independent of $\tilde{M}$, so that

$$0 = \tilde{M}\frac{dC}{d\tilde{M}} = \tilde{M}\frac{\partial C}{\partial \tilde{M}}\bigg|_{\tilde{a}} + \tilde{\beta}(\tilde{a})\frac{\partial C}{\partial \tilde{a}}. \qquad (12.56)$$

From these observations we see that

$$\mathcal{D} = \gamma_{\mathcal{O}} + \frac{\tilde{\beta}(\tilde{a})}{C} \frac{\partial C}{\partial \tilde{a}}. \qquad (12.57)$$

Thus, $\mathcal{D}$ is, in a sense, a "physicalized" version of $\gamma_{\mathcal{O}}$.

Substituting in the above formula we find

$$\mathcal{D} = -bga(1 + g_1 a + g_2 a^2 + \cdots)$$
$$+ (-b\tilde{a}^2)(1 + c\tilde{a} + \cdots) \frac{(r_1 + 2r_2\tilde{a} + \cdots)}{(1 + r_1\tilde{a} + \cdots)}. \qquad (12.58)$$

We could now expand out in terms of $\tilde{a}$, converting $a$ to $\tilde{a}$ using Eq. (12.48). Alternatively, we can eliminate $\tilde{a}$ and find the series expansion in terms of $a$. The results are more compact in the $a$ scheme:

$$\mathcal{D} = -bga(1 + r_1^{\mathcal{D}} a + r_2^{\mathcal{D}} a^2 + \cdots), \qquad (12.59)$$

with

$$r_1^{\mathcal{D}} = g_1 + r_1/g, \qquad (12.60)$$

$$r_2^{\mathcal{D}} = g_2 + \frac{1}{g} \left(2r_2 + cr_1 - r_1^2 - 2r_1(\tilde{\tau} - \tau)\right), \qquad (12.61)$$

and so on. Note that these coefficients are independent of the FS and independent of the tilde RS variables, with the explicit $g_i$ and $\tilde{\tau}, \tilde{c}_j$ dependences exactly cancelling with the implicit dependences from the $r_i$ coefficients; see Eqs. (12.19), (12.50). Thus, the $r_i^{\mathcal{D}}$ coefficients only depend, in the usual way, on the RS variables $\tau, c_j$ associated with $a$.

As usual, we can construct the $\rho_j$ invariants for the quantity $\mathcal{D}$:

$$\rho_1^{\mathcal{D}}(Q) = \tau - r_1^{\mathcal{D}}, \qquad (12.62)$$

$$\rho_2^{\mathcal{D}} = c_2 + r_2^{\mathcal{D}} - cr_1^{\mathcal{D}} - (r_1^{\mathcal{D}})^2, \qquad (12.63)$$

and these coincide with the $\sigma$'s. Indeed, it is easy to see that the "effective-charge-type" RS/FS used in the definition of the $\sigma$'s corresponds to the usual effective-charge scheme for $\mathcal{D}$, so the equivalence of $\rho_j^{\mathcal{D}}$ to $\sigma_j$ is true for all $j$.

The calculation can be straightforwardly extended to higher orders. Defining

$$\Delta \equiv \tilde{\tau} - \tau = b\ln(\tilde{M}/M), \quad s_i \equiv \frac{r_i}{g}, \tag{12.64}$$

the first three invariants are

$$\boldsymbol{\sigma}_1(Q) = \tau - g_1 - s_1, \tag{12.65}$$

$$\sigma_2 = c_2 + g_2 - g_1 c - g_1^2 + 2s_2 - 2g_1 s_1$$
$$- s_1^2(1+g) - 2s_1\Delta, \tag{12.66}$$

$$\sigma_3 = c_3 + cg_1^2 + 4g_1^3 - 6g_1 g_2 + 2g_3 - 2c_2 g_1$$
$$+ 6\tilde{c}_2 s_1 - 4cg_1 s_1 + 12g_1^2 s_1$$
$$- 6g_2 s_1 - 5cs_1^2 - 2cgs_1^2 + 12g_1 s_1^2$$
$$+ 6gg_1 s_1^2 + 4s_1^3 + 6gs_1^3 + 2g^2 s_1^3$$
$$- 6c_2 s_1 + 4cs_2 - 12g_1 s_2 - 12s_1 s_2 - 6ys_1 s_2 + 6s_3$$
$$+ (12g_1 s_1 - 10cs_1 + 12s_1^2 + 6gs_1^2 - 12s_2)\Delta + 6s_1\Delta^2. \tag{12.67}$$

Using these formulas the values of the invariants can be found from Feynman-diagram calculations performed in any convenient RS/FS.

## 12.6. The Exponentiation Theorem

The $(k+1)$th-order approximation is defined by truncating the series for $C$, $\gamma_\mathcal{O}$, and $B$. The resulting approximant, in general, will have a residual RS/FS dependence that is formally of order $a^{k+1}$. The optimization conditions correspond to requiring the RG equations to be exactly satisfied, with no remainder. (To avoid notational clutter, we leave it understood that, henceforth, any RS/FS-dependent symbol $(a, \tilde{a}, r_i, \text{etc.})$ stands for the optimized value of that quantity.)

At second order we saw that the $\tilde{\tau}$ optimization equation gave $r_1 = 0$. In third order $(k = 2)$ the $\tilde{\tau}$ equation (12.42), in which

$\partial r_2/\partial \tilde{\tau} = r_1$, reduces to

$$(1 + c\tilde{a} + \tilde{c}_2 \tilde{a}^2)(r_1 + 2r_2\tilde{a}) - r_1 = 0. \tag{12.68}$$

Also, the $\tilde{c}_2$ equation (12.43), in which the $\tilde{B}_2(\tilde{a})$ factor cancels out because $\partial r_2/\partial \tilde{c}_2 = 0$, becomes just

$$r_1 + 2r_2\tilde{a} = 0. \tag{12.69}$$

Substituting this back into the previous equation gives $r_1 = 0$. Substituting that result back into the second equation then gives $r_2 = 0$. The result generalizes to all orders.

**Theorem (Nakkagawa and Niégawa).** *The solution to the $\tilde{\tau}$ and $\tilde{c}_j$ optimization equations is*

$$r_1 = r_2 = \cdots = r_k = 0. \tag{12.70}$$

*Thus, $C = 1$ in the optimal scheme, and all perturbative corrections are effectively exponentiated and reabsorbed into the anomalous dimension $\gamma_\mathcal{O}$.*

**Proof.** The $\tilde{c}_j$ optimization equation follows from Eq. (12.43):

$$\sum_{i=j+1}^{k} \frac{\partial r_i}{\partial \tilde{c}_j} \tilde{a}^i + \tilde{a}^{j+1} \frac{\tilde{B}_j(\tilde{a})}{j-1} \frac{\partial C}{\partial \tilde{a}} = 0, \tag{12.71}$$

where $\partial C/\partial \tilde{a} = \sum_{i=1}^{k} i r_i \tilde{a}^{i-1}$. Recall that all terms up to and including $\tilde{a}^k$ must cancel in any RS, thus determining $\partial r_i/\partial \tilde{c}_j$. By starting the sum at $i = j+1$ we have already used the fact that $\partial r_i/\partial \tilde{c}_j$ must vanish for $i < j$ and for $i = j$, as noted at the end of Sec. 12.4.

We begin by considering the case $j = k$. The first term vanishes, as there are no terms in the sum, so we find that in the optimal scheme

$$\frac{\partial C}{\partial \tilde{a}} = 0. \tag{12.72}$$

Next, consider the case $j = k - 1$. In any scheme, cancellation of the $\tilde{a}^k$ terms requires

$$\frac{\partial r_k}{\partial \tilde{c}_{k-1}} = -\frac{r_1}{k-2}. \tag{12.73}$$

In the optimal scheme the left-hand side must vanish, since $\partial C/\partial \tilde{a}$ vanishes in the optimization equation (12.71). Thus, in the optimal scheme, $r_1 = 0$. Proceeding to the case $j = k - 2$ we can find $\partial r_k/\partial \tilde{c}_{k-2}$ as a sum of $r_1 c$ and $r_2$ terms. In the optimal scheme this must vanish, and since we already have $r_1 = 0$, we now find that $r_2 = 0$, too. We may then proceed to successively lower $j$ cases to see that other $r_i$'s vanish. Finally, we reach $j = 1$, where we are dealing with the $\tilde{\tau}$ equation, which gives us $r_{k-1} = 0$. Substituting back into $\partial C/\partial \tilde{a} = \sum_{i=1}^{k} i r_i \tilde{a}^{i-1} = 0$ then shows that $r_k = 0$. □

## 12.7. The Optimization Equations

The fact that $C = 1$ in the optimal scheme allows us to simplify the remaining optimization equations, which follow from Eqs. (12.44)–(12.46) with the $i$ summations truncated at $i = k$.

We first recall that the $B_j(a)$ functions are related by

$$B_j(a) = \frac{(j-1)}{a^{j-1}} B(a) I_j(a) \tag{12.74}$$

to the $I_j(a)$ integrals

$$I_j(a) \equiv \int_0^a dx \frac{x^{j-2}}{B(x)^2}. \tag{12.75}$$

(See Sec. 7.2.) The $g_j$ optimization equations involve a related set of integrals

$$J_j(a) \equiv \int_0^a dx \frac{x^{j-2}}{B(x)}, \tag{12.76}$$

while the $c_j$ optimization equations involve

$$I_{j,i}(a) \equiv (i+1) \int_0^a dx \, x^i I_j(x), \tag{12.77}$$

which can be simplified by interchanging the order of the two integrations, as follows:

$$I_{j,i}(a) = (i+1) \int_0^a dx\, x^i \int_0^x dy\, \frac{y^{j-2}}{B(y)^2}$$

$$= \int_0^a dy\, \frac{y^{j-2}}{B(y)^2} \int_y^a dx(i+1)x^i$$

$$= \int_0^a dy\, \frac{y^{j-2}}{B(y)^2} \left(a^{i+1} - y^{i+1}\right), \tag{12.78}$$

giving us

$$I_{j,i}(a) = a^{i+1} I_j(a) - I_{i+j+1}(a). \tag{12.79}$$

(This corresponds to going back to the form in Eq. (12.35) rather than Eq. (12.36); compare with Exercise 12.2.)

Thus, the $\tau$, $c_j$, and $g_j$ optimization equations can be written as

$$\sum_{i=1}^{k} \frac{\partial r_i}{\partial \tau} \tilde{a}^i - ga \sum_{i=0}^{k} g_i a^i = 0, \quad \text{``}j = 1\text{''} \tag{12.80}$$

$$\sum_{i=j}^{k} \frac{\partial r_i}{\partial c_j} \tilde{a}^i + g \sum_{i=0}^{k} g_i I_{j,i}(a) = 0, \quad j = 2, \ldots, k, \tag{12.81}$$

$$\sum_{i=j}^{k} \frac{\partial r_i}{\partial g_j} \tilde{a}^i + g J_{j+1}(a) = 0. \quad j = 1, \ldots, k. \tag{12.82}$$

In each of these equations the first term is a polynomial in $\tilde{a}$ that must precisely cancel out the terms up to and including $\tilde{a}^k$ present in the second term, if it were expanded out in a power series in $\tilde{a}$. Previously, we have used the notation $\mathbb{T}_n[G(a)] \equiv G_0 + G_1 a + \cdots + G_n a^n$. Here we will need $\tilde{\mathbb{T}}_n$ as the equivalent operation in the expansion parameter $\tilde{a}$. Thus, we may rewrite the equations (swapping the order of the two terms and dividing out a $g$ factor) as

$$a \sum_{i=0}^{k} g_i a^i - \tilde{\mathbb{T}}_k \left[ a \sum_{i=0}^{k} g_i a^i \right] = 0, \quad \text{``}j = 1\text{''} \tag{12.83}$$

$$\sum_{i=0}^{k} g_i I_{j,i}(a) - \tilde{\mathbb{T}}_k \left[ \sum_{i=0}^{k} g_i I_{j,i}(a) \right] = 0, \quad j = 2, \dots, k, \tag{12.84}$$

$$J_{j+1}(a) - \tilde{\mathbb{T}}_k \left[ J_{j+1}(a) \right] = 0. \quad j = 1, \dots, k. \tag{12.85}$$

However, note that the arguments of the $\tilde{\mathbb{T}}_k$'s are all functions of $a$, rather than $\tilde{a}$, so it is best to think of the $\tilde{\mathbb{T}}_k[G]$ operation in three stages (i) expand $G$ as series in $a$ up to $a^k$, (ii) convert $a$ to $\tilde{a}$ using Eq. (12.48):

$$a = \tilde{a}(1 + \tilde{V}_1 \tilde{a} + \tilde{V}_2 \tilde{a}^2 + \cdots), \tag{12.86}$$

(iii) re-expand as a series in $\tilde{a}$, and truncate after the $\tilde{a}^k$ term.

A further simplification results from the realization that, since $C = 1$, we do not need to know the optimized value of $\tilde{a}$; nor do we need to know the $\tilde{c}_j$'s or $\tilde{\tau}$: they do not enter into the optimized result for $F$, which just involves evaluating $\langle \mathcal{O} \rangle$ in the optimal scheme. Thus, what we need to do is to take combinations of the optimization equations in which $\tilde{a}$ and the $\tilde{V}_i$'s cancel out. From the resulting equation combinations we can solve for the $g_j$ coefficients in terms of the "principal variables" $a, c_2, \dots c_k$. (Note that the $I$ and $J$ integrals are functions of these principal variables.) Finally, we can use the invariants, $\sigma_i$ and $\boldsymbol{\sigma}_1(Q)$, and the int-$\beta$ equation to determine the optimized result. Note that when $r_i{=}0$ the $\sigma_j$'s have exactly the same form as the usual $\rho_j$ invariants with $g_i$'s in place of $r_i$'s.

> **The optimization equations** involve $I$'s up to $I_{2k+1}$ and $J$'s up to $J_{k+1}$. However, using relations between the $J$'s and $I$'s and the complete-sum identities obeyed by the $I$'s, these integrals can all be expressed in terms of $I_2$ to $I_k$. (See Appendix 9.A.)

It may be possible to find a general solution for the $g_j$ coefficients in $(k+1)$th order, akin to the results of Chapter 9. However, we will content ourselves with illustrating the above observations in the case of third order.

## 12.8. Third-Order Approximation

In third order ($k = 2$) we have four remaining optimization equations, in the variables $\tau$, $c_2$, $g_1$, and $g_2$. From Eqs. (12.83)–(12.85) these are

$$a(1 + g_1 a + g_2 a^2) - \tilde{a} - (g_1 + \tilde{V}_1)\tilde{a}^2 = 0, \qquad (12.87)$$

$$(aI_2 - I_3) + g_1(a^2 I_2 - I_4) + g_2(a^3 I_2 - I_5)$$

$$-\frac{1}{2}\tilde{a}^2 = 0, \qquad (12.88)$$

$$J_2 - \tilde{a} - \left(-\frac{c}{2} + \tilde{V}_1\right)\tilde{a}^2 = 0, \qquad (12.89)$$

$$J_3 - \tfrac{1}{2}\tilde{a}^2 = 0. \qquad (12.90)$$

Taking the $g_1$ equation minus the $\tau$ equation cancels the $\tilde{a}$ terms and, not coincidentally, the $\tilde{V}_1$ terms, leaving

$$J_2 - a(1 + g_1 a + g_2 a^2) + \left(\frac{c}{2} + g_1\right)\tilde{a}^2 = 0. \qquad (12.91)$$

An $\tilde{a}^2$ term remains, but we can substitute from the $g_2$ equation to obtain

$$J_2 + (c + 2g_1)J_3 - a(1 + g_1 a + g_2 a^2) = 0. \qquad (12.92)$$

Taking the $g_2$ equation minus the $c_2$ equation cancels the $\tilde{a}^2$ terms, giving

$$J_3 - \big((aI_2 - I_3) + g_1(a^2 I_2 - I_4) + g_2(a^3 I_2 - I_5)\big) = 0. \qquad (12.93)$$

We may solve these last two equations for $g_1, g_2$ in terms of the principal variables $a, c_2$.

**As mentioned above,** we can express all the integrals involved at $(k + 1)$th order in terms of just $I_2$ to $I_k$. In this case, then, all the integrals can be expressed in terms of $I_2$ alone. The resulting expressions are rather cumbersome, though.

From the four original equations we have extracted just two equations that give us the $g_1, g_2$ coefficients that we need. There are effectively two other equations that we can just ignore; they would determine $\tilde{a}$ and $\tilde{V}_1$ (which gives $\tilde{\tau}$ and, combined with the int-$\tilde{\beta}$ equation of the tilde scheme, would then fix $\tilde{c}_2$), but we have no need to obtain values for these variables.

To relate the principal variables to $Q$ and the invariants (whose values are to be obtained from Feynman-diagram calculations performed in any convenient scheme), we substitute the optimal-scheme quantities into the expressions for $\sigma_2$ and $\sigma_1(Q)$, combining the latter with the int-$\beta$ equation to eliminate $\tau$. In the optimal scheme, since $r_i$ (and hence $s_i \equiv r_i/g$) vanish, the formula for $\sigma_2$ reduces to

$$\sigma_2 = c_2 + g_2 - g_1 c - g_1^2, \tag{12.94}$$

which is the familiar form of a $\rho_2$ invariant, but with $g_i$'s as the coefficients. Similarly, in the optimal scheme

$$\sigma_1(Q) = \tau - g_1 = K^{(3)}(a) - g_1, \tag{12.95}$$

where $K^{(3)}(a)$ is the third-order approximation to the $K(a)$ function of Sec. 6.3. (It can also be expressed in terms of $I_2$; see Appendix 9.A.)

## 12.9. A Simpler Approach

In fact, there is a simpler approach that allows us to get directly to the equations determining the optimal $g_i$'s. Consider the physical quantity $\mathcal{D}$ defined in Eq. (12.53), which we showed is given by Eq. (12.57), so that $\mathcal{D} = \gamma_{\mathcal{O}}$ when $C = 1$. That suggests that we consider $F$ in the form:

$$F = A \exp \int_{[0]}^{a} dx \frac{\mathcal{D}(x)}{\beta(x)}, \tag{12.96}$$

where "[0]" is a shorthand for the same "lower limit of 0 with subtraction of the suitable infinite scheme-independent constant," as in Eq. (12.8). Formally, this expression for $F$ is valid quite generally, and is independent of the RS used, so it satisfies RG equations saying that the total dependences on $\tau$ and $c_j$ all vanish. What we are doing in RS/FS optimization is equivalent to a normal RS optimization applied to $F$, except that the approximants being optimized are not truncations of the perturbation series for $F$, but are approximants formed by truncating the perturbation series for $\mathcal{D}$ and $\beta$. That is, the $(k+1)$th approximant to $F$ is given by substituting

$$\hat{\mathcal{D}}(x) \equiv \frac{\mathcal{D}(x)}{(-bg)} = x \sum_{i=0}^{k} r_i^{\mathcal{D}} x^i, \quad \beta(x) = -bx^2 \sum_{j=0}^{k} c_j x^k \qquad (12.97)$$

into Eq. (12.96). The optimization equations follow from requiring the $\tau$ and $c_j$ derivatives to vanish. (Note that when we take such derivatives the infinite constant plays no role and the "[0]" lower limit can safely be replaced by 0, since the resulting integrals converge.) For $\tau$ we have

$$0 = \frac{1}{F} \frac{\partial F}{\partial \tau} = \frac{\partial a}{\partial \tau} \frac{\mathcal{D}(a)}{\beta(a)} + \int_0^a dx \left. \frac{\partial \mathcal{D}}{\partial \tau} \right|_x \frac{1}{\beta(x)}$$

$$= -g \left( \hat{\mathcal{D}}(a) - \sum_{i=1}^{k} \frac{\partial r_i^{\mathcal{D}}}{\partial \tau} J_{i+1} \right), \qquad (12.98)$$

while for $c_j$

$$0 = \frac{1}{F} \frac{\partial F}{\partial c_j} = \frac{\partial a}{\partial c_j} \frac{\mathcal{D}(a)}{\beta(a)} + \int_0^a dx \left( \left. \frac{\partial \mathcal{D}}{\partial c_j} \right|_x \frac{1}{\beta(x)} + \frac{\mathcal{D}(x)}{\beta(x)^2} bx^{j+2} \right)$$

$$= g \left( -\hat{\mathcal{D}}(a) I_j - \sum_{i=j}^{k} \frac{\partial r_i^{\mathcal{D}}}{\partial c_j} J_{i+1} + \sum_{i=0}^{k} r_i^{\mathcal{D}} I_{i+j+1} \right). \qquad (12.99)$$

Substituting the series form for $\hat{\mathcal{D}}(a)$ leads to

$$-\sum_{i=1}^{k} \frac{\partial r_i^{\mathcal{D}}}{\partial \tau} J_{i+1} + \sum_{i=0}^{k} r_i^{\mathcal{D}} a^{i+1} = 0, \qquad (12.100)$$

$$\sum_{i=j}^{k} \frac{\partial r_i^{\mathcal{D}}}{\partial c_j} J_{i+1} + \sum_{i=0}^{k} r_i^{\mathcal{D}} I_{j,i} = 0, \qquad (12.101)$$

where $I_{j,i}(a) = a^{i+1}I_j(a) - I_{i+j+1}(a)$ arises from the first and third terms of Eq. (12.99).

The derivatives $\partial r_i^{\mathcal{D}}/\partial \tau$ and $\partial r_i^{\mathcal{D}}/\partial c_j$ are the usual RS dependences of perturbative coefficients, and can be quickly found from the expressions for the $\rho_i^{\mathcal{D}}$ invariants. Thus,

$$\frac{\partial r_1^{\mathcal{D}}}{\partial \tau} = 1, \quad \frac{\partial r_2^{\mathcal{D}}}{\partial \tau} = c + 2r_1^{\mathcal{D}}, \quad \frac{\partial r_2^{\mathcal{D}}}{\partial c_2} = -1. \tag{12.102}$$

Using these results, and recalling that in the FS/RS optimal scheme the optimized $r_i^{\mathcal{D}}$'s equal the optimized $g_i$'s, the reader can quickly check that at third order $(k = 2)$ Eqs. (12.100) and (12.101) lead directly to Eqs. (12.92) and (12.93).

At fourth order $(k = 3)$ the $\tau, c_2, c_3$ equations reduce to

$$J_2 + (c + 2g_1)J_3 + (c_2 + 2cg_1 + 3g_2)J_4$$
$$-a(1 + g_1a + g_2a^2 + g_3a^3) = 0, \tag{12.103}$$

$$J_3 + 2g_1 J_4 - (I_{2,0} + q_1 I_{2,1} + g_2 I_{2,2} + g_3 I_{2,3}) = 0, \tag{12.104}$$

$$\frac{1}{2}J_4 - (I_{3,0} + g_1 I_{3,1} + g_2 I_{3,2} + g_3 I_{3,3}) = 0. \tag{12.105}$$

One can explicitly check that these are indeed the equations one would obtain from appropriate combinations of Eqs. (12.83), (12.84), (12.85).

**Note that** Eq. (12.96) can be written just as

$$F = AQ^{[\mathcal{D}]}$$

in the notation used in Sec. 2.7. The scale dimension $\mathcal{D}$ is a physical quantity that depends on a single massive variable $Q$ by the magic of dimensional transmutation, as discussed in Chapter 2. As $Q \to \infty$ one finds $F \propto (\ln Q)^{-g}$, while in the infrared limit, if $\mathcal{D}$ tends to a constant $\mathcal{D}^*$, one will have $F \propto Q^{\mathcal{D}^*}$ as $Q \to 0$ (see Appendix 12.A).

## 12.10.  Conclusions

The optimization approach to the problem of RS/FS dependence is less daunting than it appears at first sight. There are $3k$ scheme variables at $(k + 1)$th order and $k$ coefficients, $r_i$. However, $k$ of the optimization equations lead to $r_1 = \cdots = r_k = 0$, so that $C = 1$, so that another $k$ variables $(\tilde{\tau}, \tilde{c}_2, \ldots, \tilde{c}_k)$ need not be solved for. That leaves just $k$ combinations of optimization equations that can be solved for $g_1, \ldots, g_k$ in terms of the "principal variables" $a, c_2, \ldots, c_k$. In fact, these equations can be obtained more directly by the approach in the last section. By substituting in the expressions for the invariants, one can then solve for all the needed quantities. The last step will require an iterative algorithm, as in ordinary optimization (see Chapter 10).

The RS/FS optimization of $F$ described in this chapter is not the same as optimizing $\mathcal{D}$ as a perturbative physical quantity and then exponentiating the integral of $\mathcal{D}$ with respect to $dQ/Q$. However, there are strong indications that the two approaches generally give very similar results. At high energies a PWMR-like approximation to the RS/FS optimization equations gives the same result as the PWMR approximation to $\mathcal{D}$; see Exercise 12.5. In the infrared limit, if a fixed-point limit occurs, then again the results are the same as for optimizing $\mathcal{D}$; see Appendix 12.A. However, as discussed in that appendix, in the pinch-mechanism case there are interesting differences.

There are applications to various quantities, besides structure-function moments — for example, heavy quarkonium decays to hadrons, $B$ decays to charmonium, or Higgs-boson decay to hadrons. Such quantities have a factorized form involving the wavefunction at the origin or, in the last case, the quark masses. For applications involving parton distribution functions and fragmentation functions there is more work to be done. We have only considered the non-singlet case; the flavour-singlet case involves matrices describing quark–gluon mixing. Also, our analysis has used the language of structure-function moments, which is convenient theoretically since it reduces a convolution integral to a simple product. However, phenomenologically, it seems preferable to deal directly with the parton distributions using parton-evolution (DGLAP) equations. It would be valuable to see if the moments-based optimization approach can be reformulated in that language and put into practice.

## Appendix 12.A: Infrared Limit for Factorized Quantities

If a fixed-point limit occurs, the generalization of the analysis in Sec. 11.5 is straightforward. For $a$ near to $a^*$ one has $B(x) \approx \sigma(a^* - a)$, so that $I_j$ is given by Eq. (11.10) and hence, from Eq. (12.77), one finds

$$I_{j,i} \approx -(i+1)\frac{a^{*i+j-2}}{\sigma^2}\ln(a^* - a). \tag{12A.1}$$

One also has

$$J_j \approx -\frac{a^{*j-2}}{\sigma}\ln(a^* - a). \tag{12A.2}$$

At third order the optimization equations (12.92), (12.93) are potentially dominated by divergent $\ln(a^* - a)$ terms. Requiring these to vanish gives

$$1 + (c + 2g_1^*)a^* = 0, \tag{12A.3}$$

$$-\sigma a^* + (1 + 2g_1^* a^* + 3g_2^* a^{*2}) = 0, \tag{12A.4}$$

which, when solved for $g_1^*, g_2^*$, the limiting values of the coefficients of $\mathcal{D}$, yield the equivalent of Eq. (11.20) with $r$'s renamed as $g$'s. Thus, the fixed-point result for $F$ gives $F \propto Q^{\mathcal{D}^*}$, where $\mathcal{D}^*$ is exactly the same infrared limit as that obtained by optimizing $\mathcal{D}$ as a perturbative physical quantity. It may similarly be shown that the fixed-point limit of the fourth-order equations (12.103)–(12.105) leads to results for $g_1^*, g_2^*, g_3^*$ that are the equivalent of Eq. (11.25). It seems likely that this result will hold at any order.

If a pinch mechanism operates, however, things are more subtle and the result is not the same as obtained by applying ordinary optimization to $\mathcal{D}$. The $I_j$ integral is now divergent proportional to $1/\delta^3$, given by Eq. (11.32), so that, from Eq. (12.79),

$$I_{j,i} \approx (a^{\star i+1} - a_{\mathrm{p}}^{i+1})\left(\frac{a_{\mathrm{p}}^{j-2}}{\eta^2}\frac{\pi}{2\delta^3}\right) + O\left(\frac{1}{\delta}\right). \tag{12A.5}$$

The $J_j$ integrals also diverge, but only like $1/\delta$:

$$J_j \approx \int dx \frac{x^{j-2}}{\eta((x-a_{\rm p})^2 + \delta^2)} \approx \frac{a_{\rm p}^{j-2}}{\eta} \frac{\pi}{\delta}. \tag{12A.6}$$

The third-order optimization equations, (12.92), (12.93), are dominated, respectively, by $J$ terms and $I$ terms, giving

$$1 + (c + 2g_1^\star)a_{\rm p} = 0, \tag{12A.7}$$

$$(a^\star - a_{\rm p}) + g_1^\star(a^{\star 2} - a_{\rm p}^2) + g_2^\star(a^{\star 3} - a_{\rm p}^3) = 0. \tag{12A.8}$$

In the latter equation a factor of $a^\star - a_{\rm p}$ may be discarded since the pinch mechanism requires that $a^\star > a_{\rm p}$ (and we can expect the limit $a^\star \to a_{\rm p}$ to correspond to the interface with the fixed-point mechanism). Solving for $g_1^\star, g_2^\star$ yields results distinct from Eq. (11.39). As before the conditions $B(a_{\rm p}) = B'(a_{\rm p}) = 0$ lead to Eq. (11.40) so $a_{\rm p} = -2/c$ and $c_2^\star = c^2/4$. Hence,

$$g_1^\star = -\frac{c}{4}, \quad g_2^\star = -\frac{c^2}{4} \frac{(6 - ca^\star)}{(4 - 2ca^\star + c^2 a^{\star 2})}. \tag{12A.9}$$

Substituting in the definition of the $\sigma_2$ invariant gives

$$\sigma_2 = -\frac{c^2}{4} \frac{(6 - ca^\star)}{(4 - 2ca^\star + c^2 a^{\star 2})} + \frac{7}{16}c^2, \tag{12A.10}$$

which leads to a quadratic equation for $a^\star$ in terms of the invariants $c, \sigma_2$. For the pinch mechanism to operate we must have $a_{\rm p} > 0$ and $a^\star > a_{\rm p}$, which requires

$$c < 0 \quad \text{and} \quad \frac{13}{48} < \frac{\sigma_2}{c^2} < \frac{7}{16}. \tag{12A.11}$$

The lower limit is the same as in Eq. (11.43), since it is the interface with the fixed-point mechanism, but the upper limit is slightly smaller. Another difference from the ordinary case is that the result for $\hat{\mathcal{D}}^\star$:

$$\hat{\mathcal{D}}^\star = \frac{a^\star(4 - 3ca^\star)}{(4 - 2ca^\star + c^2 a^{\star 2})} \tag{12A.12}$$

means that it is restricted to the range

$$\frac{5}{3} < -c\hat{\mathcal{D}}^\star < 3. \tag{12A.13}$$

(Recall that $c$ here must be negative, and that $\mathcal{D} \equiv -bg\hat{\mathcal{D}}$.)

At fourth and higher orders we will actually need the subleading $1/\delta$ terms in the $I_j$ and $I_{j,i}$ integrals. In Eq. (11.32), expanding the $x^{j-2}$ numerator about the pinch point $a_\mathrm{p}$ gives

$$I_j \approx \frac{a_\mathrm{p}^{j-2}}{\eta^2} \int dx' \frac{1}{(x'^2 + \delta^2)^2} \left( 1 + (j-2)\frac{x'}{a_\mathrm{p}} \right.$$
$$\left. + \frac{(j-2)(j-3)}{2} \frac{x'^2}{a_\mathrm{p}^2} + \cdots \right), \tag{12A.14}$$

where $x' \equiv x - a_\mathrm{p}$ and the limits of integration may be taken as $-\infty$ to $\infty$, effectively, since we only need the terms divergent as $\delta \to 0$. Hence, we find

$$I_j \approx \frac{a_\mathrm{p}^{j-2}}{\eta^2} \frac{\pi}{2} \left( \frac{1}{\delta^3} + \frac{(j-2)(j-3)}{2a_\mathrm{p}^2} \frac{1}{\delta} \right) + O(1). \tag{12A.15}$$

Using this result and Eq. (12.79) we can then obtain, for a combination of $I_{j,i}$'s in which the leading terms cancel, that

$$I_{j+1,i} - I_{j,i}a_\mathrm{p} \approx \frac{\pi}{2\eta^2\delta} a_\mathrm{p}^{j-3}((j-2)a^{\star i+1} - (i+j-1)a_\mathrm{p}^{i+1}). \tag{12A.16}$$

At fourth order, the first optimization equation, (12.103), is dominated by $J$ terms and gives

$$1 + (c + 2g_1^\star)a_\mathrm{p} + (c_2^\star + 2cg_1^\star + 3g_2^\star)a_\mathrm{p}^2 = 0, \tag{12A.17}$$

while the other two are each dominated by $1/\delta^3$ terms whose cancellation requires

$$(a^\star - a_\mathrm{p}) + g_1^\star(a^{\star 2} - a_\mathrm{p}^2) + g_2^\star(a^{\star 3} - a_\mathrm{p}^3) + g_3^\star(a^{\star 4} - a_\mathrm{p}^4) = 0. \tag{12A.18}$$

If we take Eq. (12.105) minus $a_{\mathrm{p}}$ times Eq. (12.104) the leading divergences cancel, leaving $1/\delta$ terms whose cancellation requires

$$\eta a_{\mathrm{p}}^2(1 + 4g_1^\star a_{\mathrm{p}}) - (1 + 2g_1^\star a_{\mathrm{p}} + 3g_2^\star a_{\mathrm{p}}^2 + 4g_3^\star a_{\mathrm{p}}^3) = 0. \qquad (12\mathrm{A}.19)$$

From the pinch-point conditions $B(a_{\mathrm{p}}) = B'(a_{\mathrm{p}}) = 0$ and the $\eta$ definition as $\eta \equiv \frac{1}{2}B''(a_{\mathrm{p}})$ we have

$$c_2^\star = -\frac{(3 + 2ca_{\mathrm{p}})}{a_{\mathrm{p}}^2}, \quad c_3^\star = \frac{(2 + ca_{\mathrm{p}})}{a_{\mathrm{p}}^3}, \quad \eta = \frac{(3 + ca_{\mathrm{p}})}{a_{\mathrm{p}}^2}. \qquad (12\mathrm{A}.20)$$

From these equations and the $\sigma_2, \sigma_3$ definitions one may proceed to solve for the $g_1^\star, g_2^\star, g_3^\star$ coefficients in terms of the invariants and $a_{\mathrm{p}}$, and find a sixth-order polynomial that determines $a_{\mathrm{p}}$. Finally, the infrared limit $\mathcal{D}^\star$ can be found.

The infrared behaviour of $F$ is then $F \propto Q^{\mathcal{D}^\star}$, where the exponent is $\mathcal{D}(a)$ with $g_i = g_i^\star$ and $a = a^\star$. However, returning to Eq. (12.96) involving $\int dx\, \mathcal{D}(x)/\beta(x)$ and taking $\beta(x)$ to have the pinch-point form one would find that the exponent is given by $\mathcal{D}^\star(a)$ at $a_{\mathrm{p}}$, rather than at $a^\star$. All is well, however, because the optimization equation (12A.18), or Eq. (12A.8) in the third-order case, is precisely the condition that $\mathcal{D}^\star(a_{\mathrm{p}})$ and $\mathcal{D}^\star(a^\star)$ are the same.

**Exercise 12.1.** Give an alternative proof of the theorem that the normalization constant $A$ in Eq. (12.8) is RS invariant by suitably adapting Osborn's proof of the Celmaster–Gonsalves relation (see Sec. 6.5).

**Exercise 12.2.** Show the equivalence of Eqs. (12.36) and (12.35) for $\frac{1}{\langle \mathcal{O} \rangle}\frac{\partial \langle \mathcal{O} \rangle}{\partial c_j}$ in two ways:

(i) Integrate by parts in Eq. (12.35) and use the differential equation satisfied by the $\beta_j$ functions, Eq. (7.12).
(ii) In the right-hand side of Eq. (12.36), express $\beta_j(x)$ in terms of the integral $I_j(x)$ (see Eq. (12.75) and Sec. 7.2) to get

$$\frac{1}{\langle \mathcal{O} \rangle}\frac{\partial \langle \mathcal{O} \rangle}{\partial c_j} = -\frac{1}{b}\int_0^a dx\, I_j(x)\gamma'(x).$$

Write $I_j(x)$ as an integral over another dummy variable $y$ and then perform the $x, y$ integrations in the other order.

**Exercise 12.3.** Show that the $\tau$ RG equation (12.44) can be seen as the $j \to 1$ limit of the $c_j$ RG equation (12.45) in the same sense, "$c_1$" $\to c - (j-1)\tau$, as in Exercise 7.3. Show that the same goes for the corresponding $\tilde{\tau}$ and $\tilde{c}_j$ equations with "$\tilde{c}_1$" $\to c - (j-1)\tilde{\tau}$.

**Exercise 12.4.** Show that $F$ can be expanded out as a perturbation series

$$F = A|ca|^g (1 + r_1^F a + r_2^F a^2 + \cdots)$$

and find the first few coefficients. Calculate the invariants for $F$ (using the form for a physical quantity with the power $\mathrm{P} = g$). Show that these are related to the $\sigma_j$'s, which are the $\rho_j^{\mathcal{D}}$ invariants associated with $\mathcal{D} \equiv (Q/F)dF/dQ$ in just the way expected from the results of Exercise 7.6.

**Exercise 12.5.** Consider the equivalent of the PWMR approximation for the optimized $g_i$ coefficients in the factorized case. Show that the results, for $k = 1, 2, 3$ are the same as for the $r_i$ coefficients of the ordinary case; see Eq. (9.25). (It seems likely that this result will hold for all $k$.)

# Chapter 13

# Exploring All-Orders OPT in the Small-$b$ (BZ) Limit

## 13.1. Introduction

The constant $b$ is $(33-2n_f)/6$ in QCD, which means that asymptotic freedom is lost once $n_f$ exceeds $16\frac{1}{2}$. The constant $c$, although positive at low $n_f$, changes sign at $n_f = 8\frac{1}{19}$. Thus, as we saw in Chapter 11, when $n_f$ approaches $16\frac{1}{2}$ from below one finds fixed-point behaviour, with the infrared couplant $a^*$ proportional to $16\frac{1}{2} - n_f$. That property was noted by Caswell, and later Banks and Zaks proposed making an expansion about $n_f = 16\frac{1}{2}$. (The idea generalizes to an $SU(N)$ theory, and to other gauge theories; in some of those cases the critical $n_f$ can be an integer value.)

The BZ expansion is a distinctly unusual sort of expansion. It is not like a typical perturbative method, where a theory with some parameter $\lambda$ is exactly soluble at $\lambda = 0$ and one can step away from $\lambda = 0$ in a Taylor-series fashion. Here the theory with $n_f = 16\frac{1}{2}$ is *not* soluble; it would be a non-asymptotically-free "delicate" theory in which the $\beta$ function starts at order $a^3$ (see Exercises 6.1 and 7.7). That non-trivial special theory is not actually relevant,

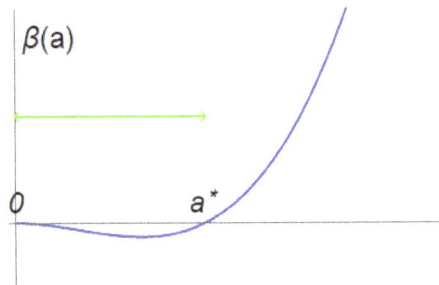

Fig. 13.1.   If the couplant at some finite energy scale lies in the region 0 to $a^*$, then it is trapped in that region at all energies. In the BZ limit $a^*$ tends to zero. (Note, however, that if $a > a^*$ then we would be, in the limit, in the non-trivial "delicate" theory.)

either. One must begin, not *at* $n_f = 16\frac{1}{2}$, but infinitesimally below that value. Also, one must assume that the couplant, at some finite energy scale, lies in the range 0 to $a^* \sim -1/c$. It is then trapped in that infinitesimal region at all energy scales. See Fig. 13.1.

The BZ expansion is normally discussed — as we shall do in the next section — only within a restricted class of "regular" renormalization schemes, where perturbative coefficients have a simple, polynomial dependence on $n_f$. However, infinitely many schemes — and in some sense most schemes — are not "regular." In particular, the "optimal" scheme is not. In "regular" schemes one needs only $k$ terms of the perturbation series to obtain $k$ terms of the BZ expansion, but in other schemes the information needed is distributed among higher-order terms. In general all orders are required. Turning that observation around, the BZ expansion can be viewed as a "playground" in which one can analytically investigate arbitrarily high orders of OPT in QCD. Admittedly, this adopts the "drunk-under-the-lamppost" principle of looking, not where we really want to, but where there is enough light to make a search. The deep and difficult issues that we would like to study – "renormalons" and factorially growing coefficients — are simply absent in the BZ limit. Nevertheless, this exploration provides some interesting insights and employs some methods that may have wider applicability.

In this chapter, we only consider quantities $\mathcal{R}$ with $\mathrm{P} = 1$.

## 13.2. BZ Expansion in "Regular" Schemes

For $n_f$ just below $16\frac{1}{2}$ the $\beta$ function has a zero at a very small $a^*$, proportional to $(16\frac{1}{2} - n_f)$. Its limiting form,

$$a_0 \equiv \frac{8}{107}b = \frac{8}{321}\left(16\frac{1}{2} - n_f\right), \qquad (13.1)$$

serves as the expansion parameter for the Banks–Zaks (BZ) expansion. To proceed, one first rewrites all perturbative coefficients, eliminating $n_f$ in favour of $a_0$. The first two $\beta$-function coefficients, which are RS invariant, become:

$$b = \frac{107}{8}a_0, \qquad (13.2)$$

$$c = -\frac{1}{a_0} + \frac{19}{4}. \qquad (13.3)$$

Note that $c$ is large and *negative* in the BZ context.

We will consider a class of "primary" physical quantities for which the $\rho_i$ invariants have the form

$$\rho_i = \frac{1}{a_0}\left(\rho_{i,-1} + \rho_{i,0}a_0 + \rho_{i,1}a_0^2 + \cdots\right). \qquad (13.4)$$

This class includes $\mathcal{R}_{e+e-}$, if we ignore the $\sum q_i$ terms. Note, however, that not all physical quantities have $\rho_i$'s of this form; the scale dimension $\mathcal{D}$ of $\mathcal{R}_{e+e-}$, for instance, would not (see Exercise 7.6). Within the class of so-called "regular" schemes, the $\beta$-function coefficients, $bc_i$, are polynomial in $n_f$, and hence polynomial in $a_0$, so that

$$c_i = \frac{1}{a_0}\left(c_{i,-1} + c_{i,0}a_0 + c_{i,1}a_0^2 + \cdots\right). \qquad (13.5)$$

Note that this equation is a property of the scheme, irrespective of the physical quantity, whereas Eq. (13.4) is a property of the physical quantity, irrespective of the scheme. For "primary" quantities in "regular" schemes we have

$$r_i = r_{i,0} + r_{i,1}a_0 + r_{i,2}a_0^2 + \cdots. \qquad (13.6)$$

**In fact,** for certain quantities the numerator of Eq. (13.4) is a polynomial whose highest term is $\rho_{i,i}a_0^{i+1}$, and in certain "rigid" schemes, such as $\overline{\text{MS}}$, the $a_0$ series for $c_i$ and $r_i$ truncate after the $c_{i,i-1}$ and $r_{i,i}$ terms. These properties are unimportant here, but are crucial in the opposite limit, the large-$b$ approximation.

Expanding in powers of $a_0$ the zero of the $\beta$ function is found to be

$$a^* = a_0(1 + (c_{2,-1} + c_{1,0})a_0 + \cdots), \qquad (13.7)$$

and hence the infrared limit of $\mathcal{R}$ is

$$\mathcal{R}^* = a_0(1 + (r_{1,0} + c_{2,-1} + c_{1,0})a_0 + \cdots). \qquad (13.8)$$

Since the BZ expansion parameter $a_0$ is RS invariant, the coefficients in the $\mathcal{R}^*$ series are RS invariant and can be written in terms of the $\rho_{i,j}$:

$$\mathcal{R}^* = a_0(1 + (\rho_{2,-1} + \rho_{1,0})a_0 + \cdots). \qquad (13.9)$$

Note, though, that $a^*$ is not a physical quantity and its $a_0$ expansion has RS-dependent coefficients.

At a finite energy $Q$ the result for $\mathcal{R}$ to $n$th order of the BZ expansion can be expressed as the solution an equation of the form

$$\rho_1(Q) = \frac{1}{\mathcal{R}} + \frac{1}{\hat{\gamma}^{*(n)}} \ln\left(1 - \frac{\mathcal{R}}{\mathcal{R}^{*(n)}}\right) + c\ln\left(|c|\,\mathcal{R}\right) \qquad (13.10)$$

for $n = 1, 2, 3$. (For $n \geq 4$ there are additional terms; see Appendix A for details.) Here $\mathcal{R}^{*(n)}$ and $\hat{\gamma}^{*(n)}$ are the $n$th-order approximations to $\mathcal{R}^*$ and $\hat{\gamma}^* \equiv \frac{\gamma^*}{b}$. As discussed in Chapter 11, the critical exponent $\gamma^*$ governs the manner in which $\mathcal{R}$ approaches $\mathcal{R}^*$ in the $Q \to 0$ limit:

$$(\mathcal{R}^* - \mathcal{R}) \propto Q^{\gamma^*}. \qquad (13.11)$$

In the present context it is safe to identify $\gamma^*$ with the slope of the $\beta$ function at the fixed point and so its BZ expansion is

$$\hat{\gamma}^* \equiv \frac{\gamma^*}{b} = a_0(1 + g_1 a_0 + g_2 a_0^2 + O(a_0^3)), \qquad (13.12)$$

where the $g_i$'s are the universal invariants of Grunberg:

$$g_1 = c_{1,0} = \rho_{1,0},$$
$$g_2 = c_{1,0}^2 - c_{2,-1}^2 - c_{3,-1} = \rho_{1,0}^2 - \rho_{2,-1}^2 - \rho_{3,-1}. \qquad (13.13)$$

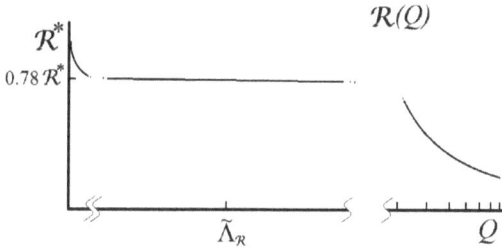

Fig. 13.2. Schematic picture of $\mathcal{R}$ as a function of $Q$ close to the BZ limit showing the three regions (i) the "spike" at very low energies, (ii) the huge flat region where the theory is "nearly scale invariant," and (iii) the slow approach to asymptotic freedom at very high energies. (Region (iii) is shown on a log scale.)

They are universal in that they do not depend on the specific physical quantity $\mathcal{R}$ being considered, and invariant because they can be expressed as combinations of the invariants $\rho_{i,j}$ (combinations in which all the $r_{i,j}$ terms cancel).

Close to the BZ limit $\mathcal{R}$ remains almost constant over a huge range of $Q$ about $\tilde{\Lambda}_{\mathcal{R}}$. This constant value is not $\mathcal{R}^*$ but $0.78\mathcal{R}^*$. More precisely, it is $\mathcal{R}^*/(1 + \chi)$ where $\ln \chi + \chi + 1 = 0$, a result that follows from Eq. (13.10) to leading order in $a_0$ with $\rho_1(Q) = 0$, corresponding to $Q = \tilde{\Lambda}_{\mathcal{R}}$. Only when $Q/\tilde{\Lambda}_{\mathcal{R}}$ becomes extremely small does $\mathcal{R}$ have a "spike" that rises up to $\mathcal{R}^*$, and only when $Q/\tilde{\Lambda}_{\mathcal{R}}$ becomes extremely large does $\mathcal{R}$ very slowly decrease to zero, as required by asymptotic freedom: See Fig. 13.2. The plateau region, where $\mathcal{R}$ stays within 10% of the value $0.78\mathcal{R}^*$, is roughly for $Q/\tilde{\Lambda}_{\mathcal{R}}$ in the range from $\exp(-0.04/a_0^2)$ to $\exp(+0.04/a_0^2)$.

Since Eq. (13.10) completely characterizes the $Q$ dependence of $\mathcal{R}$ in low-orders of the BZ expansion, it suffices to consider $\mathcal{R}^*$ and $\hat{\gamma}^*$, both of which are quantities defined in the $Q \to 0$ limit.

## 13.3. Low Orders of OPT in the BZ Limit

The procedure for obtaining the $(k + 1)$th-order OPT result for the fixed-point limit $\mathcal{R}^*$ described in Sec. 11.5 requires, of course, the numerical values of the invariants up to $\rho_k$. The great simplification in the BZ limit is that we can effectively set almost all the invariants to zero: this can be seen as follows. As $a_0 \to 0$ the most singular term in any of the $\rho_i$ is of order $1/a_0$, but each $\rho_i$ enters the analysis along with a factor of $a^{*i}$ that is of order $a_0^i$. Thus, to find the leading term in the BZ limit, we can effectively set to zero all the invariants except $c$. (Furthermore, only the $-1/a_0$ piece of $c$ will contribute.) To obtain the next-to-leading correction in $a_0$ we would also need the $\frac{19}{4}$ piece of $c$ along with the $\rho_{2,-1}/a_0$ piece of $\rho_2$ (whose value depends on the specific $\mathcal{R}$ quantity under consideration).

For $k = 2$ the OPT fixed-point results are given by Eqs. (11.20)–(11.22). In the BZ limit, we can set $\rho_2 = 0$ so that the $a^*$ equation becomes

$$(ca^* + 1)(ca^* - \tfrac{7}{3}) = 0. \tag{13.14}$$

Hence, we find $a^* = -1/c \to a_0$. The coefficients $c_2^*$, $r_1^*$, $r_2^*$ all vanish, so, in an *a posteriori* sense, the $k = 2$ OPT scheme is "regular" in the fixed-point limit. The final result for $\mathcal{R}^*$ is

$$\mathcal{R}^* = -\frac{1}{c} \to a_0. \tag{13.15}$$

Thus, exactly as in any "regular" scheme, both $a^*$ and $\mathcal{R}^*$ tend to $a_0$ in the BZ limit. The same is true for $\hat{\gamma}^*$, obtained from the slope of the optimized $\beta$ function at the fixed point.

At higher orders, though, the OPT scheme is not "regular" — the optimized $r_m^*$ coefficients, for instance, have $1/a_0^m$ pieces — and the story is more complicated. For $k = 3$ we have Eqs. (11.25)–(11.29). The $a^*$ equation in the BZ limit (where $\rho_2 = \rho_3 = 0$) reduces to

$$83 + 52ca^* = 0. \tag{13.16}$$

Thus, we do *not* get $a^* = -\frac{1}{c} \to a_0$, but $a^* \to \frac{83}{52}a_0 = 1.596a_0$. The final result for $\mathcal{R}^*$ is not $a_0$ but is $\frac{6889}{6656}a_0 = 1.035a_0$, which is remarkably close.

Results for higher orders are shown in Tables 13.1 and 13.2. The even-$k$ results are significantly better than those for odd $k$. Note that $a^*/a_0$ increases, apparently towards 4. It is perfectly acceptable for $a^*$ to differ from $a_0$, since $a^*$ is inherently scheme dependent. However, $\mathcal{R}^*$ is a physical quantity so it is reassuring that $\mathcal{R}^*/a_0$ is always close to 1. In Sec. 13.5, we will find a simple explanation for $a^*/a_0 \to 4$ and $\mathcal{R}^*/a_0 \to 1$ as $k \to \infty$.

Table 13.1. OPT results in the BZ limit for $k$ = even.

| $k$ | $\frac{a^*}{a_0}$ | $\frac{\mathcal{R}^*}{a_0}$ | $\frac{\hat{\gamma}^*}{a_0}$ |
|---|---|---|---|
| 2 | 1 | 1 | 1 |
| 4 | 1.85035 | 1.00370 | 0.9841 |
| 6 | 2.30294 | 1.00214 | 0.9742 |
| 8 | 2.58980 | 1.00137 | 0.9671 |
| 10 | 2.78928 | 1.00096 | 0.9614 |
| 12 | 2.93666 | 1.00071 | 0.9565 |
| 14 | 3.05030 | 1.00055 | 0.9523 |
| 16 | 3.14081 | 1.00043 | 0.9485 |
| 18 | 3.21470 | 1.00035 | 0.9451 |

Table 13.2. OPT results in the BZ limit for $k$ = odd.

| $k$ | $\frac{a^*}{a_0}$ | $\frac{\mathcal{R}^*}{a_0}$ | $\frac{\hat{\gamma}^*}{a_0}$ |
|---|---|---|---|
| 3 | 1.59615 | 1.03501 | 0.5602 |
| 5 | 2.17343 | 1.01119 | 0.5886 |
| 7 | 2.51313 | 1.00544 | 0.6071 |
| 9 | 2.73950 | 1.00319 | 0.6206 |
| 11 | 2.90228 | 1.00209 | 0.6311 |
| 13 | 3.02550 | 1.00147 | 0.6397 |
| 15 | 3.12231 | 1.00108 | 0.6468 |
| 17 | 3.20056 | 1.00083 | 0.6530 |
| 19 | 3.26522 | 1.00066 | 0.6583 |

The situation with $\hat{\gamma}^*$ is less clear. This is also a physical quantity (with the caveats of Appendix 11.8) so we should have $\hat{\gamma}^*/a_0 \to 1$ as $k \to \infty$. The numerical results in the tables cannot be said to support that contention, but neither are they inconsistent with it; one can make good fits to the data with functions of $k$ that very slowly approach 1 as $k = \infty$ for both even and odd $k$.

It is hard to go to much larger $k$ with the method described in this section, so we turn to an analytic approach in the next sections. Our results — albeit in approximations to OPT rather than true OPT — support the claim that $a^*/a_0 \to 4$ and that both $\mathcal{R}^*/a_0$ and

$\hat{\gamma}^*/a_0$ tend to 1 as $k \to \infty$: they also provide valuable insight into the workings of OPT at arbitrarily high orders.

## 13.4. Analytic Tools for OPT at All Orders

To make progress analytically with OPT in $(k+1)$th order it helps greatly to deal with functions and differential equations rather than with $2k$ individual $r_i$ and $c_i$ coefficients. The set of $\rho_i$ invariants — see the discussion following Eq. (7.27) — naturally follow from a single "master equation," since they are obtained by equating coefficients in

$$B_{\text{EC}}(\mathcal{R}) = \frac{a^2}{\mathcal{R}^2} \frac{\partial \mathcal{R}}{\partial a} B(a), \tag{13.17}$$

which we shall refer to as the "invariants master equation."

What we need is to formulate the $k$ optimization conditions also as a "master equation." For general $Q$ this would be a more daunting task, but in the fixed-point limit it is relatively simple — and, happily, that suffices since, as noted earlier, in the BZ limit (and for the first three terms of the BZ expansion), the entire $Q$ dependence of $\mathcal{R}$ is characterized by the two infrared quantities $\mathcal{R}^*$ and $\hat{\gamma}^*$.

We now show that the optimization conditions in the fixed-point limit, Eq. (11.16), follow from equating coefficients in the following "fixed-point OPT master equation:"

$$\frac{\partial \mathcal{R}}{\partial a} = B(a) - \frac{a}{(k-1)} \left( 2\frac{\partial B(a)}{\partial a} + \frac{B(a)}{(a^* - a)} \right). \tag{13.18}$$

(Superscripts "$(k+1)$" on $\mathcal{R}$ and $B(a)$ are omitted for brevity.) Note that $a$ here is merely a dummy variable, while $a^*$ is the optimized couplant in the infrared limit.

The first step of the proof is to note that, by the definition of $a^*$, the polynomial $B(a)$ has a factor of $a^* - a$ and can be written as

$$B(a) = \frac{(a^* - a)}{a^*} \sum_{n=0}^{k-1} \left(\frac{a}{a^*}\right)^n \hat{t}_n, \tag{13.19}$$

where $\hat{t}_n$ is a partial sum of $\beta$-function terms:

$$\hat{t}_n = \sum_{j=0}^{n} \hat{c}_j \tag{13.20}$$

with $\hat{c}_j \equiv c_j a^{*j}$. Note that $\hat{t}_n - \hat{t}_{n-1} = \hat{c}_n$ and that $\hat{t}_k = 0$ by virtue of the fixed-point condition. To show Eq. (13.19), expand the right-hand side, then use $\hat{t}_k = 0$ and define $\hat{t}_{-1} \equiv 0$ to get

$$\sum_{n=0}^{k} \left(\frac{a}{a^*}\right)^n \hat{t}_n - \sum_{n=-1}^{k-1} \left(\frac{a}{a^*}\right)^{n+1} \hat{t}_n. \tag{13.21}$$

Now put $n = n' - 1$ in the second sum and recombine the sums to get

$$\sum_{n=0}^{k} \left(\hat{t}_n - \hat{t}_{n-1}\right) \left(\frac{a}{a^*}\right)^n = \sum_{n=0}^{k} \hat{c}_n \left(\frac{a}{a^*}\right)^n = \sum_{n=0}^{k} c_n a^n, \tag{13.22}$$

which is $B(a)$, as claimed.

To prove Eq. (13.18), equate powers of $(a/a^*)^m$, using Eq. (13.19) to write $B(a)/(a^* - a)$ as a polynomial. This leads to

$$\hat{s}_m = \hat{c}_m - \frac{1}{(k-1)} \left(2m\hat{c}_m + \hat{t}_{m-1}\right). \tag{13.23}$$

Using $\hat{t}_m - \hat{t}_{m-1} = \hat{c}_m$ again and simplifying leads to the fixed-point optimization conditions, Eq. (11.16), completing the proof.

Unfortunately, Eq. (13.18) proves difficult to deal with. To make progress we have resorted to two approximations, designated PWMR and NLS. The former was discussed in Sec. 9.3 and corresponds to dropping the $O(a)$ terms in Eq. (9.25), giving

$$s_m = \frac{k - 2m}{k} c_m, \tag{13.24}$$

which is easily formulated as a "master equation":

$$\frac{\partial \mathcal{R}}{\partial a} = B(a) - \frac{2}{k} a \frac{\partial B(a)}{\partial a} \quad \text{(PWMR)}. \tag{13.25}$$

Looking at the above equation, or the original equation (13.18), it is tempting to suppose that, as $k \to \infty$, they reduce to

$$\frac{\partial \mathcal{R}}{\partial a} = B(a) \quad \text{(NLS)}. \tag{13.26}$$

We shall refer to this as the "naïve limiting scheme" (NLS). It corresponds to a well-defined RS in which $s_m = c_m$, so that the coefficients $r_m = s_m/(m+1)$ of the $\mathcal{R}$ series decrease by a factor $1/(m+1)$ relative to the coefficients of the $B$ series.

Clearly, this idea is very naïve. In the PWMR case, the actual relation only reduces to $s_m \approx c_m$ for $m \ll k$; that is, for the early part of the series only. Nevertheless, there may be a kernel of truth here, for if the series are "well behaved" the early terms should dominate. In any case, adopting this naïve idea leads us in a fruitful direction. Our investigations below will lead us to conclude that, at least in the BZ context, the NLS does yield the all-orders limit of OPT, although it is a poor guide to how fast results converge to that limit.

Using the NLS equation above to eliminate $B(a)$ in the invariants master equation (13.17) leads directly to

$$B_{\mathrm{EC}}(\mathcal{R}) = \frac{a^2}{\mathcal{R}^2} \left( \frac{\partial \mathcal{R}}{\partial a} \right)^2. \tag{13.27}$$

Taking the square root leads to

$$\frac{\partial \mathcal{R}}{\partial a} = \frac{\mathcal{R}}{a} \sqrt{B_{\mathrm{EC}}(\mathcal{R})}, \tag{13.28}$$

which is immediately integrable. (Recall that the partial derivative notation merely means that the coefficients of $\mathcal{R}$ are regarded as constant. For our purposes here it can be replaced by an ordinary derivative without creating confusion.)

The BZ limit provides us with a nice "playground" for exploring further, since it effectively corresponds to the case $B_{\mathrm{EC}}(\mathcal{R}) = 1 + c\mathcal{R}$. We continue this analysis in the next section.

## 13.5. All-Orders NLS in the BZ Limit

In the BZ limit the only one of the $\rho_n$ invariants that contributes is $c$, which is negative: $c = -1/a_0 + O(1)$ as $a_0 \to 0$. We may set $B_{\mathrm{EC}}(\mathcal{R}) = 1 + c\mathcal{R}$ in this limit. (The terms neglected can only contribute to $O(a_0)$ corrections, as argued in Sec. 13.3.) It is convenient to define

$$u \equiv \frac{-ca}{4}, \quad v \equiv -c\mathcal{R}. \tag{13.29}$$

In these variables, the NLS condition is $B = \frac{1}{4}\frac{dv}{du}$ and Eq. (13.28) becomes

$$\frac{dv}{du} = \frac{v}{u}\sqrt{1 - v}, \tag{13.30}$$

which leads to

$$\int \frac{dv}{v\sqrt{1 - v}} = \int \frac{du}{u}. \tag{13.31}$$

Performing the integral and then exponentiating both sides gives

$$\frac{1 - \sqrt{1 - v}}{1 + \sqrt{1 - v}} = u, \tag{13.32}$$

where the constant of integration has been fixed by requiring $v \to 4u$ as $u \to 0$, corresponding to the $\mathcal{R}$ series beginning $\mathcal{R} = a(1 + \cdots)$. Inverting this equation (assuming $u \leq 1$) gives

$$v = \frac{4u}{(1 + u)^2}. \tag{13.33}$$

Hence, $B = \frac{1}{4}\frac{dv}{du}$ is given by

$$B = \frac{1 - u}{(1 + u)^3}. \tag{13.34}$$

(The two formulas above are key results. They show an interesting $u \to 1/u$ duality that we shall discuss in Sec. 13.8.)

The fixed point, where $B = 0$, is at $u^* = 1$. Recalling Eq. (13.29), we see that $a^*$ is $-4/c \to 4a_0$. Nevertheless, because $u^* = 1$ in Eq. (13.33) leads to $v^* = 1$, we find $\mathcal{R}^* = -1/c \to a_0$, in agreement with the regular-scheme result.

Evaluating the slope of the $\beta$ function at the fixed point gives

$$-b\left(-\frac{4}{c}\right)u^2\frac{d}{du}\left(\frac{1-u}{(1+u)^3}\right)\Bigg|_{u=1} = \frac{-b}{2c} \to \frac{ba_0}{2}, \tag{13.35}$$

which seemingly gives $\hat{\gamma}^* \equiv \gamma^*/b = \frac{1}{2}a_0$. Here the subtlety discussed in Appendix 11.8 comes into play, since $\gamma^*$ is really the infrared limit of the "effective exponent"

$$\gamma(Q) = \frac{d\beta}{da} + \beta(a)\frac{d^2\mathcal{R}}{da^2}\Big/\frac{d\mathcal{R}}{da}. \tag{13.36}$$

Normally the second term drops out in the infrared limit because $\beta(a)$ vanishes at the fixed point. However, in the NLS the denominator $\frac{d\mathcal{R}}{da}$ also vanishes because it is $B(a) = \beta(a)/(-ba^2)$. Therefore, in the NLS case the second term contributes $-ba^2\frac{d^2\mathcal{R}}{da^2} = -ba^2\frac{dB}{da}$ which contributes an equally with the first term, thus rescaling the previous result by a factor of 2. Hence, we find $\hat{\gamma}^* = a_0$, in accord with the regular-scheme result.

The preceding discussion corresponds to the NLS result resummed to infinite order. One must now ask: Do the finite-order NLS results converge to their infinite-order form — and, if so, how fast? At $(k+1)$th order the $B$ and $v$ series are truncated, and $v^*$ is found by evaluating at $u^*$, the zero of the truncated $B$. Luckily, as with a simple geometric series, the sum of finite number of terms can be expressed fairly simply. The truncated $B$ series is

$$B^{(k+1)} = \sum_{j=0}^{k}(j+1)^2(-u)^j = \frac{1-u}{(1+u)^3} + (-1)^k k^2 \frac{u^{k+1}}{(1+u)}$$

$$\times \left(1 + O\left(\frac{1}{k}\right)\right). \tag{13.37}$$

Only for odd $k$ do we get a zero. (We will discuss even $k$ near the end of this section.) The zero of the truncated $B$ is just before $u$ reaches 1. If we put

$$u = u^* \equiv 1 - \frac{\eta(k)}{k} \tag{13.38}$$

with $\eta(k) \ll k$, we find (noting that $u^{k+1} \to e^{-\eta(k)}$) that

$$\eta(k) = 3\ln k - \ln(\ln k) - \ln(3/4) + O\left(\frac{\ln \ln k}{\ln k}\right). \tag{13.39}$$

The truncated $v$ series is

$$v^{(k+1)} = 4u\left(\sum_{j=0}^{k}(j+1)(-u)^j\right)$$

$$= 4u\left(\frac{1}{(1+u)^2} + (-1)^k k\frac{u^{k+1}}{(1+u)}\left(1+O\left(\frac{1}{k}\right)\right)\right). \tag{13.40}$$

When we substitute $u = u^*$ we find a cancellation of the $\eta(k)/k$ terms which leaves

$$v^* \approx 1 - \frac{9}{4}\frac{\ln^2 k}{k^2}. \tag{13.41}$$

This is in good accord with the numerical results in Table 13.3.

A similar analysis for $\hat{\gamma}^*$ (including the factor of 2 discussed above) leads to

$$\hat{\gamma}^* = a_0(1 + 3(-1)^{k+1}\ln k + \cdots), \tag{13.42}$$

which indicates that the NLS results for $\hat{\gamma}^*$ do *not* converge — the nominal limit of $a_0$ is "corrected" by a $\ln k$ term arising from the series-truncation effects. We indeed see this in the numerical results in Table 13.3.

Table 13.3.   NLS results in the BZ limit.

| $k$ | $4u^* = \frac{a^*}{a_0}$ | $v^* = \frac{\mathcal{R}^*}{a_0}$ | $\frac{\hat{\gamma}^*}{a_0}$ |
|---|---|---|---|
| 3 | 1.41825 | 0.69455 | 3.67 |
| 11 | 2.26825 | 0.90345 | 7.14 |
| 19 | 2.65953 | 0.95010 | 8.79 |
| 51 | 3.25059 | 0.98737 | 11.70 |
| 101 | 3.53265 | 0.99555 | 13.66 |
| 601 | 3.88410 | 0.99976 | 18.71 |

Returning to Eq. (13.37) we see that the truncated $B(u)$ function closely approximates its limiting form $\frac{1-u}{(1+u)^3}$ until $u$ gets close to 1. For odd $k$ the $(-1)^k$ "truncation effect" term causes $B$ to suddenly dive down, producing a zero. For even $k$ this term causes $B$ to suddenly shoot upwards and there is no zero. This means that there is no finite infrared limit in these orders; the "spike" in $\mathcal{R}$ goes all the way up to infinity. However, since $B$ has a minimum very close to zero the running of the couplant "almost stops" here and if we were to evaluate $v$ at this value of $u$ we would find a result close to the $\mathcal{R}^*/a_0$ obtained in the previous odd-$k$ order. A related observation is that, with only a slight change of RS, we would find an infrared limit arising from a pinch mechanism (see Appendix B).

We conclude that the NLS provides a lot of insight into OPT as $k \to \infty$, but is only a rather crude approximation to true OPT. We move on to the PWMR approximation in the next section.

## 13.6. All-Orders PWMR in the BZ Limit

As before we have $B_{\mathrm{EC}}(\mathcal{R}) = 1 + c\mathcal{R}$ in the BZ limit and we use $u \equiv \frac{-ca}{4}$ and $v \equiv -c\mathcal{R}$. In these variables, the invariants master equation (13.17) becomes

$$B = \frac{v^2}{4u^2} \frac{(1-v)}{\frac{dv}{du}},\qquad (13.43)$$

and the PWMR master equation (13.25) becomes

$$\frac{1}{4}\frac{dv}{du} = B - \frac{2}{k}u\frac{dB}{du}.\qquad (13.44)$$

We will proceed to solve these two coupled differential equations, treating $k$ as an ordinary parameter: only later we will consider the other $k$ dependence coming from the truncations of the resulting series at $(k+1)$th order. (One can explicitly check that at low $k$ this two-step approach does produce the same results as a PWMR version of the OPT procedure described in Sec. 13.3.)

We begin by making an *ansatz*:

$$B = \frac{1}{4} \frac{dv}{du} \frac{1}{\xi^2},$$ (13.45)

where $\xi$ depends on $u$. (We will actually want to view it as a function of a new variable $X$, introduced below, that itself is a function of $u$.) Substituting in Eq. (13.43) leads, in the same way as in the NLS case, to

$$\int \frac{dv}{v\sqrt{1-v}} = \int \frac{du}{u} \xi,$$ (13.46)

which leads to

$$v = \frac{4X}{(1+X)^2},$$ (13.47)

with the new variable $X$ defined by

$$X \equiv \exp \int \frac{du}{u} \xi,$$ (13.48)

or more specifically, enforcing $X \to u$ as $u \to 0$,

$$X \equiv u \exp \int_0^u \frac{d\bar{u}}{\bar{u}} (\xi - 1).$$ (13.49)

Note that

$$\frac{dX}{du} = \frac{X}{u} \xi,$$ (13.50)

so that the inverse relationship is

$$u = X \exp \int_0^X \frac{d\bar{X}}{\bar{X}} \left( \frac{1}{\xi(\bar{X})} - 1 \right).$$ (13.51)

We will now want to consider $\xi$ as a function of the new variable $X$.

We can now find $\frac{dv}{du}$ as $\frac{dv}{dX}\frac{dX}{du}$ and substitute back in the *ansatz* (13.45) to get

$$B = \frac{(1-X)}{(1+X)^3}\frac{X}{u\xi}. \tag{13.52}$$

From this we can calculate $\frac{dB}{du}$, which, after some algebra, reduces to

$$\frac{dB}{du} = \frac{B}{u}\left(\frac{(1-4X+X^2)}{(1-X^2)}\xi - 1 - X\frac{d\xi}{dX}\right). \tag{13.53}$$

Substituting this, and $\frac{1}{4}\frac{dv}{du} = \xi^2 B$ from the *ansatz* (13.45), into Eq. (13.44), leads, after cancelling a factor of $B$, to an equation for $\xi(X)$:

$$1-\xi^2 = \frac{2}{k}\left(\frac{(1-4X+X^2)}{(1-X^2)}\xi - 1 - X\frac{d\xi}{dX}\right). \tag{13.54}$$

Remarkably, this nonlinear, first-order differential equation is soluble. The trick is to write $\xi$ in the form

$$\xi = 1 - \frac{2}{k}\frac{X}{\mathcal{F}}\frac{d\mathcal{F}}{dX}. \tag{13.55}$$

This substitution, because of a cancellation of $(\mathcal{F}'/\mathcal{F})^2$ terms, leads to a *linear* second-order equation for $\mathcal{F}$. A further substitution,

$$\mathcal{F} = (1-X)^2 F, \tag{13.56}$$

leads to a Gauss hypergeometric equation, revealing that

$$F = {}_2F_1\left(-n,\frac{3}{2},-n-\frac{1}{2};X^2\right), \tag{13.57}$$

where $n \equiv k/2 - 1$. We will focus on the case of even $k$. (Curiously, the roles of odd and even $k$ are reversed relative to the NLS case.) For even $k$, the $F$ function is a polynomial of degree $n$ in $X^2$:

$$F = \frac{n!}{(2n+1)!!}\sum_{i=0}^{n}\frac{(2i+1)!!}{i!}\frac{(2(n-i)+1)!!}{(n-i)!}(X^2)^i. \tag{13.58}$$

The first few $F$'s are shown in Table 13.4. Note the "reflexive" symmetry $i \to n - i$, meaning that the coefficients are symmetric

Table 13.4. The first few $F$ polynomials and their form for large $k = 2n + 2$.

| $k$ | $n$ | $F$ |
|---|---|---|
| 2 | 0 | 1 |
| 4 | 1 | $1 + X^2$ |
| 6 | 2 | $1 + \frac{6}{5}X^2 + X^4$ |
| 8 | 3 | $1 + \frac{9}{7}X^2 + \frac{9}{7}X^4 + X^6$ |
| 10 | 4 | $1 + \frac{4}{3}X^2 + \frac{10}{7}X^4 + \frac{4}{3}X^6 + X^8$ |
| $\infty$ | $\infty$ | $(1 - X^2)^{-3/2}$ $(X \neq 1)$ |
| | | $\sqrt{n^3} \frac{\sqrt{\pi}}{2} \frac{e^{-x} I_1(x)}{x}$ $(X = 1 - \frac{x}{n})$ |

about the middle. In the $n \to \infty$ limit $F$ approaches $(1 - X^2)^{-3/2}$, except near $X = 1$, where its behaviour involves a modified Bessel function $I_1$ (see Table 13.4).

To find $u$ in terms of $X$ it is helpful to use another representation of $\xi$, namely

$$\frac{1}{\xi} = 1 - \frac{1}{n+2} \frac{X}{\mathcal{P}} \frac{d\mathcal{P}}{dX}, \qquad (13.59)$$

so that Eq. (13.51) will immediately lead to

$$u = X \mathcal{P}^{-\frac{1}{(n+2)}}. \qquad (13.60)$$

Substituting the above form for $\frac{1}{\xi}$ into the $\xi$ equation (13.54) leads again to a linear equation. One can verify that this equation is satisfied by setting

$$\mathcal{P} = (1 + X)^4 P \qquad (13.61)$$

with

$$P = \frac{1}{(n+1)} \frac{1}{(1+X)} \left( [n+1 - (n-1)X] F - 2(1 - X)X^2 \frac{dF}{d(X^2)} \right). \qquad (13.62)$$

Table 13.5.   The first few $P$ polynomials and their form for large $k = 2n + 2$.

| $k$ | $n$ | $P$ |
|-----|-----|-----|
| 2 | 0 | $1$ |
| 4 | 1 | $1 - X + X^2$ |
| 6 | 2 | $1 - \frac{4}{3}X + \frac{26}{15}X^2 - \frac{4}{3}X^3 + X^4$ |
| 8 | 3 | $1 - \frac{3}{2}X + \frac{15}{7}X^2 - \frac{15}{7}X^3 - \frac{15}{7}X^4 - \frac{3}{2}X^5 + X^6$ |
| 10 | 4 | $1 - \frac{8}{5}X + \frac{12}{5}X^2 - \frac{8}{3}X^3 + \frac{62}{21}X^4 - \frac{8}{3}X^5 + \frac{12}{5}X^6 - \frac{8}{5}X^7 + X^8$ |
| $\infty$ | $\infty$ | $(1 - X)^{-1/2}(1 + X)^{-5/2}$    $(X \neq 1)$ |
| | | $\sqrt{n}\frac{\sqrt{\pi}}{4} e^{-x} I_0(x)$    $(X = 1 - \frac{x}{n})$ |

The numerator turns out to have a $(1 + X)$ factor, so that $P$ is a polynomial of degree $2n$ in $X$. The first few $P$'s are shown in Table 13.5. These polynomials also have a "reflexive" property.

Yet another expression for $\zeta$ is

$$\xi = \frac{(1 + X)}{(1 - X)} \frac{P}{F}, \tag{13.63}$$

which can be proved by substituting for $P$ and simplifying to reach Eq. (13.55). Using this form of $\xi$ in Eq. (13.52) gives

$$B = (1 - X)^2 F P^{-\left(\frac{n+1}{n+2}\right)}. \tag{13.64}$$

As noted in the tables, both $F$ and $P$ polynomials have simple limits as $k \to \infty$, provided that $X \neq 1$. It is easy to see that $X \to u$ and that all formulas revert to their NLS forms in this limit. Thus, it is clear that $v^*$ must ultimately tend to 1, so that $\mathcal{R}^* = a_0$ in accord with the BZ limit.

However, to go further analytically and determine how fast the finite-order PWMR results approach their infinite-order form is beset with difficulties; the subtleties when $X \sim 1$ are crucial. The theory of hypergeometric functions when two parameters go to infinity is formidably complicated. Moreover, in any finite order we need to re-express both $B$ and $v$ as series, not in $X$ but in $u$; then find $u^*$ from the zero of the truncated $B$ series; and then evaluate the truncated $v$ series at $u = u^*$. Nevertheless, we can explore these issues numerically with Mathematica. We have been able to explore up to $k \approx 100$ and the numerical results are presented in Table 13.6. It appears that $v^*$ approaches 1 significantly faster than in the NLS case:

$$v^* \sim 1 - A\frac{\ln k/k_0}{k^2}, \qquad (13.65)$$

with $A \approx 0.08$ and $k_0 \approx 2.5$, roughly.

The ratio of $v$ to its NLS form $v_{\rm NLS} \equiv \frac{4u}{(1+u)^2}$ stays very close to 1 in the entire relevant range $0 < u < u^*$, although it strongly deviates thereafter. See Fig. 13.3

Table 13.6.   PWMR results in the BZ limit.

| $k$ | $4u^* = \frac{a^*}{a_0}$ | $v^* = \frac{R^*}{a_0}$ | $\hat{\gamma}^*/a_0$ |
|---|---|---|---|
| 2 | 1 | 1 | 1 |
| 4 | 1.56878 | 0.99743 | 1.0526 |
| 10 | 2.41100 | 0.99893 | 1.1064 |
| 18 | 2.88641 | 0.99952 | 1.1371 |
| 50 | 3.46514 | 0.99990 | 1.1869 |
| 100 | 3.69257 | 0.99997 | 1.2183 |

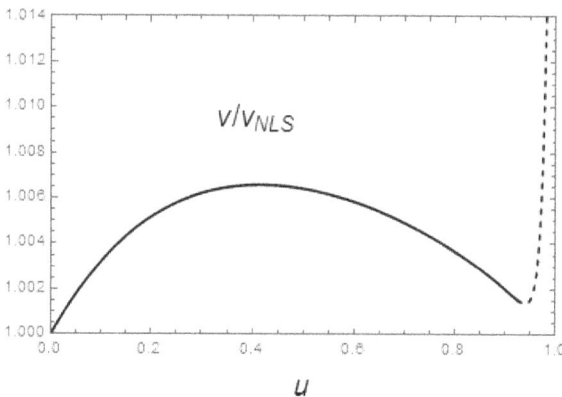

Fig. 13.3.   Plot of $v$ divided by $v_{\rm NLS} \equiv \frac{4u}{(1+u)^2}$ as a function of $u$ for PWMR at $k = 100$. The curve is shown dashed beyond $u = u^* = 0.92314$.

The $v$ series is also much better behaved than in NLS, where the magnitude of the coefficients increased in arithmetic progression: $v_{\mathrm{NLS}} = 4u \sum_j (j+1)(-u)^j$. In the PWMR case, the coefficients $v_j$ in

$$v = 4u \sum_{j=0}^{k} v_j(-u)^j \qquad (13.66)$$

are plotted in Fig. 13.4 for $k = 100$. The initial $(j+1)$ growth is suppressed by a more-than-exponential decay (a crude fit is $(j+1)\exp(-0.019j^{3/2})$). The middle coefficient $j = \frac{k}{2}$ is exactly zero because of the $k - 2j$ factor in the PWMR relation between $s_j$ and $c_j$ coefficients, Eq. (13.24). The coefficients remain very small thereafter. The somewhat bad behaviour of the last few coefficients is almost entirely suppressed by the $u^j$ factor, even at $u = u^*$, the largest relevant $u$, and it actually plays a beneficial role. This can be seen in Fig. 13.5 which plots the partial sums of $n_{\max}$ terms of the $v$ series, Eq. (13.66), at $u = u^*$ in the case $k = 100$. The series has pretty well converged after 50 terms, but including 25 more terms significantly reduces the error. The very last term makes an

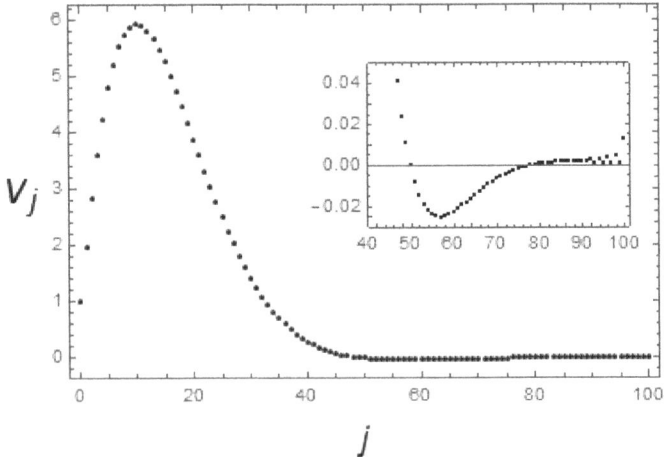

Fig. 13.4.   Coefficients $v_j$ in the series expansion of $v(u) = 4u \sum_{j=0}^{k} v_j(-u)^j$, for PWMR with $k = 100$. The inset shows the higher-order coefficients on a finer scale.

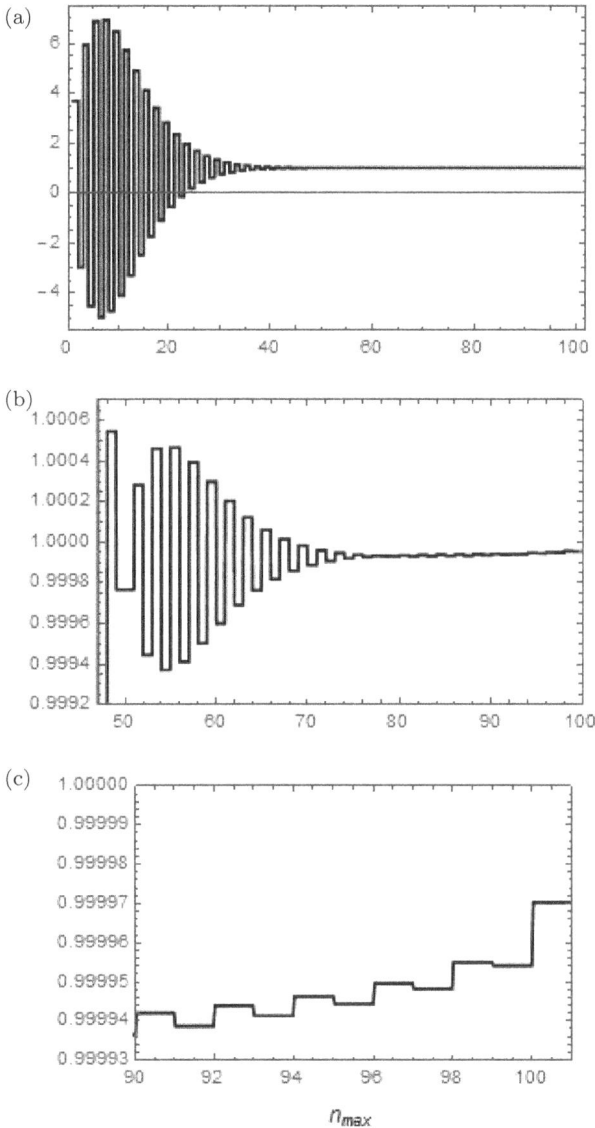

Fig. 13.5.   The partial sums $4u \sum_{j=0}^{n_{\max}} v_j(-u^*)^j$ versus $n_{\max}$ for the $v^*$ series in the case $k = 100$. The plots use three different scales, so as to show that (a) the series has crudely converged after 50 terms but (b) a slight adjustment from 50 to 75 terms reduces the error quite significantly, and (c) the last term makes an unexpectedly large change, given the trend of the preceding terms, but this further improves the result and means that the last term is, within a factor of 2, a good measure of the actual error.

unexpectedly large correction, but this further reduces the error and means that the last term provides quite a realistic error estimate.

The series for $\hat{\gamma}^*$, which is just $d\beta/da|^*$, is much worse behaved. Also the sequence of results for $\hat{\gamma}^*$ in Table 13.6 appear to diverge, though at a much slower rate than in NLS. It is reasonable to hope that the extra subtleties in full OPT would lead to $\hat{\gamma}^*$ converging to $a_0$, albeit very, very slowly, in view of the low-order OPT results in Tables 13.1 and 13.2.

We have not been able to extend the analysis to the full fixed-point master equation, (13.18). One can get to an equation similar to Eq. (13.54), but with an extra term involving $u/(u - u^*)$ that seems intractable. Moreover, the parameter $u^*$ can only be fixed after the $B(u)$ function is found, and expressed as a truncated series, so the interaction between analytic subtleties and truncation effects is even more complicated and delicate.

## 13.7.  BZ Expansion in All-Orders OPT

Setting aside the difficult issue of how fast results converge as $k \to \infty$, the results of the last section confirm that the simple NLS formulas from Sec. 13.5,

$$v = \frac{4u}{(1 + u)^2}, \tag{13.67}$$

$$B = \frac{1 - u}{(1 + u)^3}, \tag{13.68}$$

represent the all-orders limit of PWMR — and presumably of true OPT too — in the BZ limit. As previously noted, these formulas give the same BZ limit for $\mathcal{R}^*$ and $\hat{\gamma}^*$ as "regular" schemes. We now show that higher terms in the BZ expansion are reproduced correctly by all-orders NLS.

Before discussing the general proof it is instructive to look at next-to-leading order in the BZ expansion. At this level, we now need two of the invariants, $c$ and $\rho_2$ so we take

$$B_{\mathrm{EC}} = 1 + c\mathcal{R} + \rho_2 \mathcal{R}^2. \tag{13.69}$$

(In fact, only the $\rho_{2,-1}$ piece of $\rho_2$ would contribute when we re-expand the results in powers of $a_0$. However, it will not be necessary to carry out that step explicitly, since once we show equivalence to the EC scheme, a "regular" scheme, we are bound to get the same BZ expansion to the corresponding order in $a_0$.) Recall that the NLS condition and the invariants master equation together lead to Eq. (13.28),

$$\frac{d\mathcal{R}}{da} = \frac{\mathcal{R}}{a}\sqrt{B_{\text{EC}}(\mathcal{R})}, \tag{13.70}$$

which now gives

$$\int \frac{d\mathcal{R}}{\mathcal{R}\sqrt{1 + c\mathcal{R} + \rho_2 \mathcal{R}^2}} = \int \frac{da}{a}. \tag{13.71}$$

Integration yields

$$\ln\left(\frac{4\mathcal{R}}{2 + c\mathcal{R} + 2\sqrt{1 + c\mathcal{R} + \rho_2 \mathcal{R}^2}}\right) = \ln a, \tag{13.72}$$

where the constant of integration has been fixed so that $\mathcal{R} \to a$ as $a \to 0$. One can now exponentiate and solve for $\mathcal{R}$, and then $B(a)$ can be found from $d\mathcal{R}/da$. As before we define $u = -ca/4$ and $v = -c\mathcal{R}$. The zero of $B$ is at

$$u^* = \frac{1}{\sqrt{1 - 4\frac{\rho_2}{c^2}}}, \tag{13.73}$$

and in terms of these variables we find

$$v = \frac{4u}{(1 + 2u + \frac{u^2}{u^{*2}})}, \tag{13.74}$$

$$B = \frac{1 - \frac{u^2}{u^{*2}}}{(1 + 2u + \frac{u^2}{u^{*2}})^2}. \tag{13.75}$$

It is now straightforward to check that $v$ evaluated at $u = u^*$ gives

$$\mathcal{R}^* = -\frac{v^*}{c} = -\frac{c}{2\rho_2}\left(1 - \sqrt{1 - \frac{4\rho_2}{c^2}}\right), \tag{13.76}$$

which is the root of $B_{\mathrm{EC}}(\mathcal{R}) = 0$. Thus, the $\mathcal{R}^*$ of all-orders NLS agrees with the $\mathcal{R}^*$ of the EC scheme. Also, $\hat{\gamma}^*$, defined as the infrared limit of Eq. (13.36), which leads to

$$\hat{\gamma}^* = -2a^2 \left.\frac{dB}{da}\right|^*, \qquad (13.77)$$

with the factor-of-2 subtlety as in Sec. 13.5, can be shown to reduce to

$$\hat{\gamma}^* = -\mathcal{R}^2 \left.\frac{dB_{\mathrm{EC}}}{d\mathcal{R}}\right|^*, \qquad (13.78)$$

which is the $\hat{\gamma}^*$ of the EC scheme.

The general proof is really just a special case of the general formal arguments that $\mathcal{R}^*$ and $\hat{\gamma}^*$ (properly defined) are invariant under RS transformations. From Eq. (13.70) we can see immediately that $B(a)$, equal to $d\mathcal{R}/da$ in NLS, must vanish when $B_{\mathrm{EC}}$ vanishes; thus the $\mathcal{R}$ evaluated at $a = a^*$ in NLS must agree with the $\mathcal{R}^*$ defined as the zero of the EC $\beta$ function. Furthermore, the equivalence of the two equations for $\hat{\gamma}^*$ above can be proved just from the NLS condition $B = d\mathcal{R}/da$ and Eq. (13.70), without assuming any specific form for $B_{\mathrm{EC}}$.

## 13.8.  $a \rightarrow a^{*2}/a$ Duality

It is easily verified that under $u \rightarrow u^{*2}/u$ the $v$ of Eq. (13.74) remains invariant, while the $B$ of Eq. (13.75) transforms to $-(u^2/u^{*2})B$. These properties are even easier to spot in Eqs. (13.67), (13.68), in the BZ-limit case, where $u^* = 1$.

Let us try to trace the origin of these properties. Consider a transformation

$$a \longrightarrow \frac{\lambda^2}{a}, \qquad (13.79)$$

with some positive constant $\lambda$. We postulate that $\mathcal{R}$ and all the $\rho_i$ invariants remain invariant and that the $\beta$-function equation,

$\mu \frac{da}{d\mu} = \beta(a)$ maintains its form. The latter condition means that

$$\frac{da}{d\tau} = -a^2 B(a), \tag{13.80}$$

where $\tau = b \ln(\mu/\tilde{\Lambda})$, must transform to

$$\frac{d}{d\tau}\left(\frac{\lambda^2}{a}\right) = -\left(\frac{\lambda^2}{a}\right)^2 B^{\mathrm{T}}(a), \tag{13.81}$$

where $B^{\mathrm{T}}(a) \equiv B(\frac{\lambda^2}{a})$. This requires

$$B^{\mathrm{T}}(a) = -\frac{a^2}{\lambda^2} B(a). \tag{13.82}$$

If $B(a)$ vanishes at $a = a^*$ then $B^{\mathrm{T}}(a)$ must too. Thus $\lambda^2/a^*$ must be a zero of $B(a)$. If we assume that there is only one zero, then we must take $\lambda = a^*$.

The transformation of $\frac{d\mathcal{R}}{da}$ would be

$$\frac{d\mathcal{R}}{da} \longrightarrow \frac{d\mathcal{R}}{d(\frac{\lambda^2}{a})} = -\frac{a^2}{\lambda^2}\frac{d\mathcal{R}}{da}. \tag{13.83}$$

Note that this is the same transformation rule as for $B$ above. Thus, the NLS scheme-fixing condition, $\frac{d\mathcal{R}}{da} = B(a)$, transforms into itself. It is straightforward to check that the same is true of the invariants master equation (13.17). It thus seems that an $a \to a^{*2}/a$ duality is not special to the BZ limit, but is a general property of all-orders NLS and hence of all-orders OPT.

## 13.9. Conclusions

While BZ results are most simply obtained in a restrictive class of "regular" schemes, the same results emerge from "irregular" schemes, though they then require consideration of all orders of perturbation theory. Results in OPT for the fixed-point value $\mathcal{R}^*$ are never far from the BZ result and converge quite nicely to it. The error at $(k+1)$th order shrinks as $\ln^2 k/k^2$ in NLS, as $\ln k/k^2$ in PWMR, and probably slightly faster in true OPT. These results provide some insight into how the subtle features of OPT conspire to improve finite-order results.

Of course, in the BZ limit the EC scheme, or any "regular" scheme, is clearly *better* than OPT; their results converge *immediately* to the right result. The BZ limit is a case where we have an extra piece of information — and the general principle that we should make use of all available information takes precedence here over the Principle of Minimal Sensitivity. One should keep in mind, though, that the BZ limit, $n_f \rightarrow 16\frac{1}{2}$, is not a physical one. It is an open question whether or not OPT gives better results than the EC scheme for $n_f = 16$, the closest physical case.

The situation with the critical exponent $\gamma^*$ is much less satisfactory. While the all-orders NLS formulas produce the correct result, the finite-order NLS and PWMR results do not actually converge. In true OPT the results might converge but, if so, the convergence is extremely slow. The problem may stem from trying to obtain $\gamma^*$ as a by-product of the optimization of $\mathcal{R}^*$. If one is principally interested in $\gamma^*$ itself, then one should construct its own perturbation series and optimize that. However, our reason here for studying $\gamma^*$ was not for its own sake, but as a shortcut to obtaining $\mathcal{R}(Q)$ at non-zero $Q$, relying on Eq. (13.10), which holds for the first three orders of the BZ expansion. That was very convenient because we only needed the optimization conditions at the fixed point, and these are analytically much simpler than for general $Q$. However, the natural procedure is to optimize $\mathcal{R}(Q)$ itself. There is no reason to suppose that the convergence of OPT for $\mathcal{R}(Q)$ at non-zero $Q$ is significantly worse than for $\mathcal{R}^*$; indeed, as $Q$ gets larger we expect convergence to become much better. Thus, our difficulties with $\gamma^*$ are probably a technical, mathematical issue, rather than a problem of physical concern.

A key result is the "fixed-point OPT master equation" (13.18) which opens a route to an analytical treatment of arbitrarily high orders of OPT, given knowledge of the $\rho_i$ invariants — although here we have only been able to make progress in two simplifying approximations, NLS and PWMR. It appears that the simple NLS approximation does yield the all-orders limit of OPT, although it is a poor guide to the rate of approach to that limit. The NLS formulas, (13.33), (13.34) at leading order in the BZ expansion, and (13.74), (13.75) at next-to-leading order, are remarkably simple. They illustrate a general $a \to a^{*2}/a$ duality property of all-orders OPT that is intriguing and deserves further study.

## Appendix 13.A: BZ Expansion at General $Q$

We first examine the BZ limit. Here, the first two terms of the $\beta$ function dominate, so we may use the int-$\beta$ equation in the second-order form

$$b \ln(\mu/\tilde{\Lambda}) = \frac{1}{a} + c \ln \left| \frac{ca}{1 + ca} \right|. \tag{13A.1}$$

Combining this equation with the definition of $\rho_1(Q)$ yields the latter in terms of $a$. Recalling that in the BZ limit $c \sim -1/a_0$, $\mathcal{R}^* \sim a^* \sim a_0$, and $\mathcal{R} \sim a$, the limiting form is

$$\rho_1(Q) = \frac{1}{\mathcal{R}} + \frac{1}{\mathcal{R}^*} \ln \left( \frac{\mathcal{R}^* - \mathcal{R}}{\mathcal{R}} \right). \tag{13A.2}$$

Inverting this equation (numerically) gives $\mathcal{R}$ as a function of $\rho_1(Q) = b \ln(Q/\tilde{\Lambda}_{\mathcal{R}})$, and hence as a function of $Q$. This is the function sketched in Fig. 13.2. It is universal, at least within the class of "primary" physical quantities.

To go beyond this limiting form and develop a systematic BZ expansion for $\mathcal{R}$ at a general $Q$ is not a completely unambiguous matter. $\mathcal{R}(Q)$ is not expressible as a simple power series in $a_0$, so some thought is required in deciding how precisely to define the $n$th-order approximant. The important point, as with any approximation, is to reconcile and make best use of all available information. Simply taking the int-$\beta$ equation and then expanding it in powers of $a_0$ produces correction terms with $(a_0 - \mathcal{R})$ denominators. Higher orders bring in ever more singular terms. However, these terms simply arise from an expansion of $\ln(\mathcal{R}^* - \mathcal{R})$, reflecting the fact that the fixed point $\mathcal{R}^*$ does not stay at $a_0$, but is itself a series in $a_0$. Therefore, it is sensible to organize the expansion to reflect this fact.

Thus, before performing the integration of $1/\hat{\beta}(x)$, where $\hat{\beta}(x) \equiv \beta(x)/b$, one should first rewrite it as

$$\frac{1}{-x^2(1 + cx + c_2 x^2 + \cdots)} = \frac{-a^*}{x^2(a^* - x)P(x)}, \qquad (13A.3)$$

ensuring that the pole is in the right place, and then express it in partial fractions as

$$\frac{1}{\hat{\beta}(x)} = -\frac{1}{x^2} + \frac{c}{x} - \frac{1}{\hat{\gamma}^*(a^* - x)} + H(x). \qquad (13A.4)$$

The coefficients of the first three terms are determined by the $x \to 0$ and $x \to a^*$ limits. Hence $\hat{\gamma}^*$ is $\gamma^*/b$, where $\gamma^*$ is the slope of the $\beta$ function at the fixed point. The remainder term can be expanded as a power series, $H(x) = H_0 + H_1 x + \cdots$.

In $n$th order of the BZ expansion one may truncate the $\beta$ function after $n+1$ terms. In that case $H(x)$ is initially of the form $Q(x)/P(x)$, where $P(x)$ and $Q(x)$ are polynomials of degree $n - 1$ and $n - 2$, respectively. (For $n = 1$, $Q(x)$ vanishes.) The coefficients of $P(x)$ are of order unity as $a_0 \to 0$. The coefficients of $Q(x)$ are of order $a_0$ because of cancellations that make both $c + 1/\hat{\gamma}^*$ and $a^*/\hat{\gamma}^* - 1$ of order $a_0$. For instance, $H_0 = a_0(c_{4,-1} + 2c_{2,-1}c_{3,-1} + c_{2,-1}^3) + O(a_0^2)$. In the BZ expansion to $n$th order ($n \geq 4$) one needs coefficients up to $H_{n-4}$: for the first three orders one can drop $H(x)$ altogether.

It is now straightforward to integrate $1/\hat{\beta}(x)$ in the form of Eq. (13A.4) and hence obtain $\rho_1(Q)$ in terms of $a$ and $a^*$. One may then eliminate $a$ and $a^*$ in favour of $\mathcal{R}$ and $\mathcal{R}^*$, working to the appropriate order. The last step can be short circuited by noting that the final result must be RS invariant, and so, without loss of generality, one may choose to work in the EC scheme in which $\mathcal{R} = a$. Thus, the result to $n$th order in the BZ expansion can be expressed as

$$\rho_1(Q) = \frac{1}{\mathcal{R}} + \frac{1}{\hat{\gamma}^{*(n)}} \ln\left(1 - \frac{\mathcal{R}}{\mathcal{R}^{*(n)}}\right) + c\ln\left(|c|\,\mathcal{R}\right) + \sum_{i=0}^{n-4} \frac{H_i^{\text{EC}}\mathcal{R}^{i+1}}{i+1},$$
$$(13A.5)$$

where $\mathcal{R}^{*(n)}$ and $\hat{\gamma}^{*(n)}$ are the $n$th-order approximations to $\mathcal{R}^*$ and $\hat{\gamma}$, respectively. For small $\mathcal{R}$ (i.e., at large $Q$) this formula will agree with $(n+1)$th-order perturbation theory to the appropriate order in $\mathcal{R}$ and $a_0$.

## Appendix 13.B: Pinch Mechanism Infrared Limit

As discussed in Sec. 11.6, a finite infrared limit in OPT can occur through a pinch mechanism: The evolving $B(a)$ function of the optimized scheme develops a minimum that "pinches" the horizontal axis at a "pinch point" $a_\mathrm{p}$, which ultimately becomes a double zero of $B(a)$. The infrared limit of the couplant, however, is at an "unfixed point" $a^\star > a_\mathrm{p}$ that is *not* a zero of the $\beta$ function. The approach to $Q = 0$ involves $1/(\ln Q)^2$, and so is characterized by $\gamma^* = 0$.

In the BZ limit, $n_f \to 16\frac{1}{2}$, the pinch mechanism does not seem to occur in true OPT, at least as far as we have been able to explore it in Sec. 13.3. However, the mechanism is very close to being relevant because in the BZ limit the critical exponent $\gamma^* \sim ba_0$ tends to zero. A small or zero $\gamma^*$ gives rise to a sharp infrared "spike" in $\mathcal{R}$ plotted versus $Q$, as in Fig. 13.2 or Fig. 11.4

The NLS and PWMR approximations to OPT seem to have fixed points only in every other order (for odd $k$ in NLS, and even $k$ in PWMR). In these orders, as discussed in Sec. 13.5, the $B(u)$ function closely approximates its limiting form $(1-u)/(1+u)^3$ until $u$ gets close to 1, when it suddenly dives down, producing a zero. In the alternating orders $B(u)$ suddenly shoots upwards and there is no zero. However, $B(u)$ then has a minimum very close to the horizontal axis, so only a slight modification of the scheme would produce a "pinch point."

We first show that, in circumstances where the pinch mechanism *does* govern the infrared limit of OPT, the master equation that replaces Eq. (13.18) is

$$\frac{d\mathcal{R}}{da} = \left(\frac{1-a/a^\star}{1-a/a_\mathrm{p}}\right)\left[B(a) - \frac{a}{(k-1)}\left(2\frac{dB(a)}{da} + \frac{B(a)}{(a_p - a)}\right)\right].$$

$$(13\mathrm{B}.1)$$

(Superscripts "$(k + 1)$" on $\mathcal{R}$ and $B(a)$ are omitted for brevity.) Except for the prefactor, and the fact that $a_\mathrm{p}$ (not $a^\star$) replaces $a^*$ in the last term, this equation is identical to (13.18).

The derivation starts from Eq. (11.33) for the $s_m$ coefficients in infrared limit and uses a dummy variable $a$ to form the function

$$S(a) = \frac{d\mathcal{R}}{da} = \sum_{m=0}^{k} s_m a^m. \tag{13B.2}$$

Reorganizing the resulting double summation over $m$ and $j$ so that the latter becomes the outer summation, the inner summations become finite geometric series or derivatives thereof. The outer $j$ summation then produces terms that are $B(a)$ or $dB/da$ or $B(a_\mathrm{p})$ or $dB/da\big|_{a=a_\mathrm{p}}$. The last two vanish in the infrared limit since $a_\mathrm{p}$ is then a double zero of the $B(a)$ function. After some further algebraic tidying up the result reduces to Eq. (13B.1) above.

Note that the naïve large-$k$ limit of Eq. (13B.1) is not the NLS condition (13.26) but

$$\frac{d\mathcal{R}}{da} = \left(\frac{1 - a/a^\star}{1 - a/a_\mathrm{p}}\right) B(a) \qquad \text{(NLS}'). \tag{13B.3}$$

If we proceed in parallel with the analysis in Sec. 13.5 we find, instead of Eq. (13.31),

$$\int \frac{dv}{v\sqrt{1-v}} = \int \frac{du}{u} \sqrt{\frac{1 - u/u^\star}{1 - u/u_\mathrm{p}}}. \tag{13B.4}$$

Note that the above equations correspond to the *ansatz* form used in the PWMR analysis of Sec. 13.6 with $\xi$ replaced by

$$\xi \to \sqrt{\frac{1 - u/u^\star}{1 - u/u_\mathrm{p}}}. \tag{13B.5}$$

Doing the integrations, exponentiating both sides, and solving for $v$ leads to

$$v = \frac{4U}{(1+U)^2}, \tag{13B.6}$$

where

$$
U = \left( \frac{4u^\star u_{\mathrm{p}}}{u^\star - u_{\mathrm{p}}} \right) \left( \frac{\sqrt{\frac{1-u/u^\star}{1-u/u_{\mathrm{p}}}} - 1}{\sqrt{\frac{1-u/u^\star}{1-u/u_{\mathrm{p}}}} + 1} \right) \left( \frac{\sqrt{u^\star - u} + \sqrt{u_{\mathrm{p}} - u}}{\sqrt{u^\star} + \sqrt{u_{\mathrm{p}}}} \right)^{2\sqrt{\frac{u_{\mathrm{p}}}{u^\star}}}.
$$

$$(13\text{B}.7)$$

Note that when $u > u_{\mathrm{p}}$ (which is relevant since $u$ ranges from 0 to $u^\star$, which must exceed $u_{\mathrm{p}}$) this formula for $U$ develops an imaginary part. However, recall that both $v$ and $B$,

$$
B = \frac{(1-U)}{(1+U)^3} \frac{U}{u} \sqrt{\frac{1 - u/u_{\mathrm{p}}}{1 - u/u^\star}}
$$

$$(13\text{B}.8)$$

(cf. Eq. (13.52)), have to be expanded as series in $u$ and then *truncated* after $k$ terms, making them inevitably real.

These formulas are hard to deal with, even at low orders, especially since $u_{\mathrm{p}}$ and $u^\star$ have to be determined by the requirements that the truncated $B$ and its derivative vanish at the pinch point $u_{\mathrm{p}}$. For $k = 2, 4$ there does not seem to be any viable solution, but for sufficiently large $k$ it appears there is. Anticipating that both $u_{\mathrm{p}}$ and $u^\star$ will tend to 1 as $k \to \infty$, we define

$$
\delta \equiv \frac{1}{u_{\mathrm{p}}} - \frac{1}{u^\star}
$$

$$(13\text{B}.9)$$

and proceed to expand to lowest non-trivial order in $\delta$. This gives

$$
U \approx u \left( 1 - \frac{\delta}{2} \ln(1 - u) \right),
$$

$$(13\text{B}.10)$$

$$
v \approx \frac{4u}{(1+u)^2} - 2\delta u \frac{(1-u)}{(1+u)^3} \ln(1 - u),
$$

$$(13\text{B}.11)$$

and

$$
B \approx \frac{1-u}{(1+u)^3} - \frac{\delta}{2} \left( \frac{u}{(1+u)^3} + \frac{(1 - 4u + u^2)}{(1+u)^4} \ln(1 - u) \right).
$$

$$(13\text{B}.12)$$

Table 13.7.   NLS$'$ results, to lowest-order in $\delta$, in the BZ limit.

| $k$ | $u_{\mathrm{p}}$ | $u^{\star}$ | $\delta$ | $v^{\star} = \frac{\mathcal{R}^{\star}}{a_0}$ |
|---|---|---|---|---|
| 100 | 0.95018 | 0.97735 | 0.02925 | 0.71485 |
| 600 | 0.98819 | 0.99292 | 0.00482 | 0.95982 |
| 10,000 | 0.99895 | 0.99924 | 0.00029 | 0.99856 |

Remarkably, one can find analytic expressions for the truncated-series versions of $v$ and $B$ and thereby explore numerical results up to very high $k$ values. These results (see Table 13.7) show that indeed there is a valid solution (with $u^{\star} > u_{\mathrm{p}}$) with $\delta$ tending to zero as $\delta \sim (2/\ln 2)(1/k)$ and $\mathcal{R}^{\star}/a_0$ tending to 1.

# Bibliography

## Relevant Works of the Author

[1] *Dimensional analysis in field theory*, Ann. Phys. (NY) **132**, 383 (1981).

[2] *Resolution of the renormalisation-scheme ambiguity in perturbative QCD*, Phys. Lett. B **100**, 61 (1981).

[3] *Optimized perturbation theory*, Phys. Rev. **D 23**, 2916 (1981).

[4] *Optimized perturbation theory in the Gross–Neveu model*, Phys. Rev. D **24**, 1622 (1981).

[5] *The renormalization-scheme-dependence problem and its solution*, in *Perturbative Quantum Chromodynamics*, Tallahassee 1981, edited by D. W. Duke and J. F. Owens (AIP, New York, 1981).

[6] *Sense and nonsense in the renormalization-scheme-dependence problem*, Nucl. Phys. B **203**, 472 (1982).

[7] *Comment on "An alternative implementation of the 'principle of minimal sensitivity'"*, Phys. Rev. D **27**, 1968 (1983).

[8] *Scale-scheme ambiguities in the Brodsky–Lepage–Mackenzie procedure*, Phys. Lett. B **125**, 493 (1983) (with W. Celmaster).

[9] *An introduction to the renormalization-scheme ambiguity of perturbation theory*, published in *Radiative Corrections in* $SU(2)_L \times U(1)$, edited by B. W. Lynn and J. F. Wheater (World Scientific Press, Singapore, 1984).

[10] *Optimization and the ultimate convergence of QCD perturbation theory*, Nucl. Phys. **B 231**, 65 (1984).

[11] *Renormalization-scheme ambiguity and perturbation theory near a fixed point*, Phys. Rev. D **29**, 1682 (1984) (with J. Kubo and S. Sakakibara).

[12] *Explicit formula for the renormalization-scheme invariants of perturbation theory*, Phys. Rev. D **33**, 3130 (1986).

[13] *Optimized perturbation theory applied to factorization scheme dependence*, Nucl. Phys. **B 277**, 758 (1986) (with H. D. Politzer).

[14] *QCD perturbation theory at low energies*, Phys. Rev. Lett. **69**, 1320 (1992) (with A. C Mattingly).

[15] *Optimization of $R_{e^+e^-}$ and "freezing" of the QCD couplant at low energies*, Phys. Rev. **D 49**, 437 (1994) (with A. C. Mattingly).

[16] *The $16\frac{1}{2} - N_f$ expansion and the infrared fixed point in perturbative QCD*, Phys. Lett. B **331**, 187 (1994).

[17] *Optimization of QCD perturbation theory: results for $R_{e^+e^-}$ at fourth order*, Nucl. Phys. **B 868**, 38 (2013).

[18] *Fixed and unfixed points: infrared limits in optimized QCD perturbation theory*, Nucl. Phys. **B 875**, 63 (2013).

[19] *The effective exponent $\gamma(Q)$ and the slope of the $\beta$ function*, Phys. Lett. B **761**, 428 (2016).

[20] *Exploring arbitrarily high orders of optimized perturbation theory in QCD with $n_f \to 16\frac{1}{2}$*, Nucl. Phys. B **910**, 469 (2016).

[21] *The Banks–Zaks expansion in perturbative QCD: An update*, Mod. Phys. Lett. A **31**, 1650226 (2016).

[22] *Optimization for factorized quantities in perturbative QCD*, Nucl. Phys. B **944**, 114635 (2019).

## Feynman-Diagram Results

The Feynman-diagram calculations in QCD cited below were each, in their time, heroic accomplishments using state-of-the-art techniques and took great patience, care, and skill to achieve — and no little courage to undertake. The first five references below are for calculations of the $\beta$-function coefficients $b, c, c_2^{\overline{\text{MS}}}, c_3^{\overline{\text{MS}}}, c_4^{\overline{\text{MS}}}$, respectively. The last three references are for the $\mathcal{R}_{e^+e^-}$ coefficients $r_1^{\overline{\text{MS}}}, r_2^{\overline{\text{MS}}}, r_3^{\overline{\text{MS}}}$, respectively.

[23] H. D. Politzer, Phys. Rev. Lett. **30**, 1346 (1973); D. J. Gross and F. Wilczek, *ibid.* **30**, 1343 (1973).

[24] D. R. T. Jones, Nucl. Phys. B **75**, 531 (1974); W. Caswell, Phys. Rev. Lett. **33**, 244 (1974); E. S. Egorian and O. V. Tarasov, Theor. Mat. Fiz. **41**, 26 (1979).

[25] O. V. Tarasov, A. A. Vladimirov and A. Yu. Zharkov, Phys. Lett. B **93**, 429 (1980); S. A. Larin and J. A. M. Vermaseren, Phys. Lett. B **303**, 334 (1993).

[26] T. van Ritbergen, J. A. M. Vermaseren and S. A. Larin, Phys. Lett. B **400**, 379 (1997).

[27] P. A. Baikov, K. G. Chetyrkin and J. H. Kühn, Phys. Rev. Lett. **118**, 082002 (2017); F. Hertzog, B. Ruijl, T. Ueda, J.A. M. Vermaseren and A. Vogt, JHEP **1702**, 090 (2017); T. Luthe, A. Maier, P. Marquard and Y. Schröder, JHEP **1607**, 127 (2016); JHEP **2017**, 166 (2017).

[28] K. G. Chetyrkin, A. L. Kataev and F. V. Tkachov, Phys. Lett. B **85**, 277 (1979); M. Dine and J. Sapirstein, Phys. Rev. Lett. **43**, 668 (1979); W. Celmaster and R. J. Gonsalves, Phys. Rev. D **21**, 3112 (1980).

[29] L. R. Surguladze and M. A. Samuel, Phys. Rev. Lett. **66**, 560 (1991); *ibid* 2416 (E); S. G. Gorishny, A. L. Kataev and S. A. Larin, Phys. Lett. B **259**, 144 (1991).

[30] P. A. Baikov, K. G. Chetyrkin, J. H. Kühn and J. Rittinger, Phys. Lett. B **714**, 62 (2012); P. A. Baikov, K. G. Chetyrkin and J. H. Kühn, Phys. Rev. Lett. **101**, 012002 (2008).

## Renormalization Group

Stueckelberg and Peterman [31], and Gell Mann and Low [32], are usually cited jointly as the origin of the Renormalization Group (RG). From our perspective, the importance of the latter is that it puts the idea to use — extending perturbation theory in QED to high energies — while the special virtue of Stueckelberg–Peterman lies in the recognition that a new arbitrary parameter enters in each successive order — so that the arbitrariness is not just renormalization-scale dependence. Bogoliubov and Shirkov [33] further developed these early ideas. For an enlightening history see Fraser [34].

Wilson [35], of course, enormously broadened the RG concept and expanded the applications far beyond perturbation theory and renormalizable quantum field theories.

Especially relevant to our discussion here are the works of Callan and Symanzik [36], Weinberg [37], Coleman and E. Weinberg [38] and the classic reviews by Coleman [39], Politzer [40], Gross [41], and Peterman [42].

[31] E. C. G. Stueckelberg and A. Peterman, Helv. Phys. Acta **26**, 449 (1953).

[32] M. Gell Mann and F. Low, Phys. Rev. **95**, 1300 (1954).

[33] N. N. Bogoliubov and D. Shirkov, *Introduction to the Theory of Quantized Fields* (Wiley-Interscience, New York, 1959).

[34] J. D. Fraser, Stud. Hist. Phil. Sci. **89**, 114 (2021).

[35] K. Wilson, Phys. Rev. D **3**, 1818 (1971); Phys. Rev. B **4**, 3174 (1971); *ibid.* 3184; Rev. Mod. Phys. **47**, 773 (1975); K. Wilson and J. Kogut, Phys. Rep. **12C**, 75 (1974).

[36] C. G. Callan, Phys. Rev. D **2**, 1541 (1970); K. Symanzik, Commun. Math. Phys. **18**, 227 (1970).

[37] S. Weinberg, Phys. Rev. D **8**, 3497 (1973).

[38] S. Coleman and E. Weinberg, Phys. Rev. D **7**, 1888 (1973).

[39] S. Coleman in *Properties of the Fundamental Interactions*, edited by A. Zichichi (Editrice Compositori, Bologna, 1973).

[40] H. D. Politzer, Phys. Rep. **14C**, 129 (1974).

[41] D. J. Gross in *Methods in Field Theory*, edited by R. Balian and J. Zinn-Justin (North-Holland, Amsterdam, 1976).

[42] A. Peterman, Phys. Rep. **53C**, 157 (1979).

## Some Other Applications of PMS

We give here a few references to applications of PMS that lie outside the scope of this book. In particular, there are very interesting developments in QCD that involve RS optimization in conjunction with the introduction, *à la* CK, of a variational mass parameter [43–45].

Some other applications are to solid-state physics [46]; the effective average action method [47]; the calculation of critical exponents for $O(N)$ models [48]; calculation of accurate eigenvalues for non-uniform strings and membranes [49]; and to finite-temperature field theory [50].

A wide-ranging review, touching on many areas of physics, is by Yukalov [51], one of the pioneers in the study of approximation methods and the optimization of various kinds of perturbation-theory methods. It provides an extensive collection of references to the literature.

[43] J.-L. Kneur and A. Neveu, Phys. Rev. D **81**, 125012 (2010); *ibid* **85**, 014005 (2012); *ibid* **88**, 074025 (2013); *ibid* **92**, 074027 (2015); J.-L. Kneur, M. B. Pinto and T. E. Restrepo, Phys. Rev. D **100**, 114006 (2019).

[44] F. Siringo, Nucl. Phys. B **907**, 572 (2016); Phys. Rev. D **94**, 114036 (2016).

[45] M. Peláez, U. Reinosa, J. Serreau, M. Tissier and N. Wschebor, Rep. Prog. Phys. **84**, 124202 (2021).

[46] K. Chen and K. Haule, Nat. Commun. **10**, 3725 (2019).

[47] L. Canet, B. Delamotte, D. Mouhanna and J. Vidal, Phys. Rev. D **67**, 065004 (2003).

[48] G. De Polsi, I. Balog, M. Tissier and N. Wschebor, Phys. Rev. E **101**, 042113 (2020).

[49] P. Amore, F.M. Fernandez and M. Rodriguez, Cent. Eur. J. Phys. **10**, 913 (2012).
[50] S. Chiku and T. Hatsuda, Phys. Rev. D **58**, 076001 (1998).
[51] V. I. Yukalov, Phys. Part. Nucl. **50**, 141 (2019).

## Other References for Part I

**Chapter 2:** This chapter is based on the exposition in Ref. [1], but the content is drawn from the classic literature on the Renormalization Group cited above. "Dimensional transmutation" was named by Coleman and E. Weinberg [38]. Appendix 2.B refers to the Callan–Symanzik equations [36] and the Weinberg version with a running mass [37].

**Chapter 3:** An account of renormalization from a similar viewpoint, but with the calculational details for QED at one loop given explicitly, can be found in a chapter of Ref. [52]. A classic textbook account of renormalization in QED, including the proof of renormalizability to all orders, is in Bjorken and Drell [53]. There are many excellent modern textbooks on quantum field theory which the reader may consult for explanations of Feynman diagram calculations and their renormalization in QCD. For example, see [54, 55]

The existence of the RS-dependence problem, and a realization of its seriousness in QCD, emerged in a number of calculations in the early days of QCD; see references cited in Ref. [3]. Especially influential was the work of Celmaster and Gonsalves [56], and Celmaster and Sivers [57].

The analysis of 't Hooft followed in Exercise 3.2 is from Ref. [58].

**Chapter 4:** References for the CK expansion are [59, 60]. A related method, the Halliday–Suranyi expansion [61], has interesting similarities and differences to the CK expansion. It uses the *square* of a harmonic-oscillator operator as its $H_0$. At first order it gives exactly the same result as CK, namely $^{(0)}\langle n|H|n\rangle^{(0)}$. The HS series converges for any finite, non-zero $\Omega$, but PMS optimization is still important for getting the best out of finite orders.

The CK expansion was rediscovered by a different route in the context of $\phi^4$ field theory, and is known there as the "linear $\delta$ expansion" [62].

Highly accurate numerical results for the AHO energy levels were calculated by Hioe *et al.* [63].

Our discussion of the calculation of wavefunctions in the CK method follows [64]. See also [65].

The Gaussian effective potential (GEP) idea has a long history. Credit is due to, among others, Schiff, Rosen, Kuti, Chang, Barnes and Ghandour, Bardeen and Moshe. References can be found in [66], which discusses the concept in both quantum mechanics and in $\phi^4$ field theory.

Other examples of non-invariant approximations, and related situations, can be found in Yukalov's review [51].

For references to FAC in the RS-dependence problem, see references for Chapter 7 below.

**Chapter 5:** The large-order behaviour of the usual perturbation series for the AHO was established by Bender and Wu [67].

The discussion of high orders in the CK expansion is based on Caswell's work [59]. For the proof of convergence see [68–70] and references therein.

The "toy model" (inspired by an example in Ref. [57]) is from [10]. See also [71]. The notion of "induced convergence" is related to the "order-dependent mappings" of [72] and to the ideas of Yukalov [51].

For Laplace's method see [73]. The result, Eq. (5.36), can be found in [74].

[52] F. Halzen and A. D. Martin, *Quarks and Leptons* (Wiley, New York, 1982)

[53] J. D. Bjorken and S. D. Drell, *Relativistic Quantum Fields* ( McGraw-Hill, 1965).

[54] M. E. Peskin and D. V. Schroeder, *An Introduction to Quantum Field Theory* (CRC Press, 1995).

[55] S. Weinberg, *The Quantum Theory of Fields*, Vols. 1,2,3 (Cambridge University Press, 2005).

[56] W. Celmaster and R. J. Gonsalves, Phys. Rev. D **20**, 1420 (1979); Phys. Rev. Lett. **42**, 1435 (1979); Phys. Rev.D **21**, 3112 (1980); Phys. Rev. Lett. **44**, 560 (1980).

[57] W. Celmaster and D. Sivers, Phys. Rev. D **23**, 227 (1981).

[58] G. 't Hooft, Nucl. Phys. B **61**, 455 (1973).

[59] W. E. Caswell, Ann. Phys. (NY) **123**, 153 (1979).

[60] J. Killingbeck, J. Phys. A **14**, 1005 (1981); E. J. Austin and J. Killingbeck, J. Phys. A **15**, L443 (1982).

[61] I. G. Halliday and P. Suranyi, Phys. Lett. B **85**, 421 (1979); Phys. Rev. D **21**, 1529 (1980).

[62] A. Duncan and M. Moshe, Phys. Lett. B **215**, 352 (1988); H. F. Jones, Nucl. Phys. B (Proc. Suppl.) **16**, 592 (1990); A. Duncan and H. F. Jones, Phys. Rev. D **47**, 2560 (1993); M. Pinto and R. Ramos, Phys. Rev. D **60**, 105005 (1999).

[63] F. T. Hioe, D. MacMillen and E. W. Montroll, Phys. Rep. **43C**, 305 (1978).

[64] S. K. Kauffmann and S. M. Perez, J. Phys. A **17**, 2027 (1984); *ibid* **19**, 3807 (1986); *ibid* **20**, 2645 (1987).

[65] T. Hatsuda, T. Kunihiro and T. Tanaka, Phys. Rev. Lett. **78**, 3229 (1997).

[66] P. M. Stevenson, Phys. Rev. D **30**, 1712 (1984); Phys. Rev. D **32**, 1389 (1985).

[67] C. M. Bender and T. T. Wu, Phys. Rev. **184**, 1231 (1969); Phys. Rev. D **7**, 1620 (1973).

[68] C. M. Bender, A. Duncan and H. F. Jones, Phys. Rev. D **49**, 4219 (1994).

[69] R. Guida, K. Konishi and H. Suzuki, Ann. Phys. **241**, 152 (1995); *ibid* **249**, 109 (1996).

[70] H. Kleinert and V. Schulte-Frohlinde, *Critical Properties of $\phi^4$ Theories* (World Scientific, Singapore, 2001), Chap. 19.

[71] K. Van Acoleyen and H. Verschelde, Phys. Rev. D **69**, 125006 (2004).

[72] R. Seznec and J. Zinn-Justin, J. Math. Phys. **20**, 1398 (1979)

[73] F. W. J. Olver, *Asymptotics and Special Functions* (Academic Press, New York, 1974), Chap. 3.

[74] M. Abramowitz and I. A. Stegun, *Handbook of Mathematical Functions* (Dover, 1965), formula 6.5.34 with 6.5.11.

## Other References for Part II

**Chapter 6:** The proof that the second $\beta$-function coefficient is RS invariant, and the clear realization that the higher coefficients are not, is due to 't Hooft [75]. The RP with $c_j = 0$ ($j \geq 2$) was used, cleverly and appropriately, by 't Hooft in Ref. [76] to investigate the singularities of Green's functions in the ultraviolet limit.

For the (non-)issue of "gauge-dependence" see Appendix B of [3] and the references cited there.

The $\tilde{\Lambda}$ definition is from [3]. The conventional $\Lambda$ definition originates with Buras *et al.* [77]. The ambiguity created by solving the int-$\beta$ equation by a series in "$1/\ln(\mu/\Lambda)$" was noted by Monsay and Rosenzweig [78].

The crucial CG relation was proved by Celmaster and Gonsalves [56]. We are indebted to H. Osborn (private communication) for the instructive alternative proof, inspired by a perceptive remark of P. Landshoff at a seminar given by the author.

For QCD pressure see [79].

**Chapter 7:** Parametrization of RS dependence by $\tau$ and the $c_j$'s originates in [3]. Some streamlining of the argument is due to Politzer [93]. Note that the definition of $\rho_j$ invariants in [3] and used in [11, 15] was slightly different. Explicit expressions for the invariants can be found in Refs. [12, 80]. For the proof of Eq. (7.34) see [12].

References for the FAC/EC scheme are [81–83]. The similarity of FAC and PMS results at low orders is discussed in Refs. [84, 85].

**Chapter 8:** Based mainly on [3, 15]. For more figures illustrating the dependence of perturbative approximants on the RS parameters, see [86].

**Chapter 9:** The solution for the optimized $r_i$ coefficients is from [17]. For the PWMR approximation, see [87] and also [7].

**Chapter 10:** These examples are from Ref. [17]. See references under the "Feynman–Diagram Results" heading above. For issues involved in the actual comparison with experiment see Ref. [15].

Examples of optimization applied to QED are considered briefly in [3] (Subsection VI.b) and more thoroughly in [88]. At low energies optimization makes only a slight improvement, because the conventional "on-shell" scheme is then close to optimal, but at higher energies optimization is certainly needed.

[75]  G. 't Hooft, Lecture at 1977 Coral Gables Conf., in *Deeper Pathways in High-Energy Physics* (Orbis Scientiae, 1977) edited by B. Kursunoglu, A. Perlmutter and L. F. Scott (Plenum, New York, 1977).
[76]  G. 't Hooft, Lecture at 1977 Erice School, in *The Whys of Subnuclear Physics*, edited by A. Zichichi (Plenum, New York, 1979).
[77]  A. J. Buras, E. G. Floratos, D. A. Ross and C. T. Sachrajda, Nucl. Phys. **B 131**, 308 (1977); W. A. Bardeen, A. J. Buras, D. W. Duke and T. Muta, Phys. Rev. D **18**, 3998 (1978).
[78]  E. Monsay and C. Rosenzweig, Phys. Rev. D **23**, 1217 (1981).
[79]  A. D. Linde, Phys. Lett. B **96**, 289 (1980); K. Kajantie, M. Laine, K. Rummukainen and Y. Schroder, Phys. Rev. D **67**, 105008 (2003); G. Jackson and A. Peshier, Phys. Rev. D **95**, 054021 (2017).
[80]  C. J. Maxwell, Phys. Rev. D **29**, 2884 (1984).
[81]  G. Grunberg, Phys. Lett. **B 95**, 70 (1980); Phys. Rev. D **29**, 2315 (1984).
[82]  A. Dhar, Phys. Lett. **B 128**, 407 (1983); A. Dhar and V. Gupta, Phys. Rev. D **29**, 2822 (1984).
[83]  C. J. Maxwell, Phys. Lett. B **409**, 450 (1997); arXiv:hep-ph/9809270 (1998); C. J. Maxwell and A. Mirjalili, Nucl. Phys. B **577**, 209 (2000).
[84]  J. Kubo and S. Sakakibara, Phys. Rev. D **26**, 3656 (1982).
[85]  D. W. Duke and R. G. Roberts, Phys. Rep. **120**, 275 (1985).

[86] P. A. Rączka, Nucl. Phys. Proc. Suppl. **55 C**, 403 (1997); P. Posolda, J. Phys. G **41**, 095007 (2014).

[87] M. R. Pennington, Phys. Rev. D **26**, 2048 (1982); J. C. Wrigley, Phys. Rev. D **27**, 1965 (1983); J. A. Mignaco and I. Roditi, Phys. Lett. B **126**, 481 (1983).

[88] J. Kubo and S. Sakakibara, Z. Phys. C **14**, 345 (1982).

## Other References for Part III

**Chapter 11:** This chapter is based largely on [18], which builds upon [11, 14, 15]. For fixed point lore see the reviews cited under the "Renormalization Group" heading above, or any modern quantum-field-theory textbook.

The ideas referred to in the third paragraph are exemplified by Bloom–Gilman duality [89] and Poggio–Quinn–Weinberg (PQW) smearing [90]. For the "new methods," see Refs. [43–45] above.

Appendix 11.A is based on [19]. The issue of whether $\dot{\beta}^*$ is truly an invariant was raised by Chýla [91]. (Note that some other remarks in that paper were predicated upon the incorrect result for the $\overline{\text{MS}}$ $r_2$ coefficient discussed below.)

The fixed-point limit of OPT was first considered in [11]. At that time the $\overline{\text{MS}}$ $c_2$ coefficient was known, but no $r_2$ coefficient had been calculated. Later, the erroneous 1988 result for $r_2$ appeared to indicate no fixed point (see also the note in Sec. 10.3). However, the correct 1991 result [29] leads to freezing behaviour in OPT, as first noted, somewhat sceptically, in [92] and later, more optimistically, in [14, 15]. See the latter references for a discussion of the phenomenological consequences and a comparison with experimental data, after applying PQW smearing [90].

**Chapter 12:** The application of the PMS criterion to the problem of factorization-scheme dependence has had a rather unfortunate history. The pioneering work by Politzer [93] was marred by a trivial algebraic error, seemingly showing that the optimization equations had no solution. The error was belatedly corrected in Ref. [13]. However, Ref. [13] is, in retrospect, insufficiently general beyond second order. The formulation of Nakkagawa and Niégawa (NN) in a series of papers [94], though unduly complicated and creating unnecessary difficulties, arrived at optimization equations that are equivalent to those derived here. NN also discovered the important result that we call the *exponentiation theorem*. (Note that in all the papers just cited the coefficient "$b$" has the opposite sign to ours.)

This chapter follows our recent treatment in Ref. [22]. (See Appendix A of that reference for a detailed discussion of the relation to NN's work.)

In a fixed scheme, the $n$th moment has coefficients with $\ln^2 n$ and $\ln n$ terms, making perturbative results unreliable when $n$ is large. Reference [95] shows that "optimization" cures this problem — much as optimizing $\Omega$ in the CK expansion produces good results for the high-$n$ eigenvalues, as well as for the ground state.

**Chapter 13:** Based largely on [20]. References for the BZ expansion include [96–98]; our treatment follows [16, 21].

[89]  E. D. Bloom and F. Gilman, Phys. Rev. D **4**, 2901 (1971).
[90]  E. C. Poggio, H. R. Quinn and S. Weinberg, Phys. Rev. D **13**, 1958 (1976).
[91]  J. Chýla, Phys. Rev. D 38, 3845 (1988).
[92]  J. Chýla, A. Kataev and S. A. Larin, Phys. Lett. B **267**, 269 (1991).
[93]  H. D. Politzer, Nucl. Phys. B **194**, 493 (1982).
[94]  H. Nakkagawa and A. Niégawa, Phys. Lett. B **119**, 415 (1982); Prog. Theor. Phys. **70**, 511 (1983); *ibid* **71**, 339 (1984); *ibid* **71**, 816 (1984).
[95]  H. Nakkagawa, A. Niégawa and H. Yokota, Phys. Rev. D **34**, 244 (1986).
[96]  T. Banks and A. Zaks, Nucl. Phys. B **196**, 189 (1982).
[97]  A. R. White, Phys. Rev. D **29**, 1435 (1984); Int. J. Mod. Phys. A **8**, 4755 (1993).
[98]  G. Grunberg, Phys. Rev. D **46**, 2228 (1992).

* 9 7 8 9 8 1 1 2 5 5 6 8 7 *